NANOTECHNOLOGY SCIENCE AND TECHNOLOGY

TiO_2 NANOPARTICLES BY FLAME CVD METHOD

SYNTHESIS, CHARACTERIZATION, SIMULATION AND APPLICATION

NANOTECHNOLOGY SCIENCE AND TECHNOLOGY

Additional books in this series can be found on Nova's website under the Series tab.

Additional E-books in this series can be found on Nova's website under the E-book tab.

NANOTECHNOLOGY SCIENCE AND TECHNOLOGY

TiO_2 NANOPARTICLES BY FLAME CVD METHOD

SYNTHESIS, CHARACTERIZATION, SIMULATION AND APPLICATION

HONGYONG XIE
AND
SHI CHEN

Nova Science Publishers, Inc.
New York

LIBRARY OF CONGRESS CATALOGING-IN-PUBLICATION DATA

TiO2 nanoparticles by flame CVD method : synthesis, characterization, simulation and application / editors, Hongyong Xie, Shi Chen.

p. cm.

Includes bibliographical references and index.

ISBN 978-1-61470-262-7 (softcover : alk. paper) 1. Titanium dioxide. 2. Nanoparticles. I. Xie, Hongyong. II. Chen, Shi.

QD181.T6T55 2011

620'.5--dc23

2011022479

Published by Nova Science Publishers, Inc. + New York

CONTENTS

PREFACE

The scientific and technological issues of nanostructured particles and materials are currently attracting considerable attention. The reason for this is that nanostructured particles and materials, and the physical or chemical combination of substances at the nanometer or subnanometer scale, can lead to innovative materials with improved or even unexpected properties. Applications which could have a great impact can be anticipated in fields like catalysis, technical ceramics, membrane technology, optoelectronics, and solid-state ionics with emphasis on systems for clean energy conversion and storage. However, progress in these fields will largely depend on the pace of advance of the fundamental research on nanostructured particles and materials in solid-state chemistry, solid-state physics, and materials science.

In principle, the synthesis routes pursued today for the generation of nanostructured and nanocomposite particles and materials are based either on special solution processing techniques which have been derived from established solution chemistry, or on gas-phase deposition methods.

In solution processing, the possible combinations of constituents for the synthesis of mainly oxides' materials is restricted. Deposition from a vapor-phase mixture of the components through gas-phase techniques is more versatile, and widely applicable, as gas-phase techniques can be applied to nearly any material.

The gas-phase synthesis route makes it possible to generate new nanoparticles and nanostructured new materials from, in principle, a nearly unlimited variety of starting materials. The starting point of the gas-phase synthesis of nanostructured materials is the generation of gas-born nano-sized particles. Such a particle system represents an aerosol. Over the past decade, aerosols occurring in other disciplines have been studied in detail, especially the formation of these gas-born nano-particles, their characterization, as well as their handling. Aerosol researchers, in particular, are now applying this aerosol expertise to the field of nanostructured materials development. Also, materials scientists have begun to use aerosol formation processes for the creation of materials with improved or new properties, i.e. for the "synthesis of material properties".

Flame aerosol technology refers to the formation of fine particles from gases in flames. This technology has been practiced since prehistoric times as depicted with paintings on cave walls and in Chinese ink artwork. Today, flame technology is employed routinely in large scale manufacture of carbon blacks and ceramic commodities such as fumed silica and pigmentary titania and, to a lesser extent, for specialty chemicals such as zinc oxide and alumina powders. These powders find most of their applications as pigments and reinforcing agents. Today, the production volume of this industry is of the order of several million metric tons per year. Industries that employ flame technologies for manufacture of ceramic powders include several of the leading chemical firms from all over the world: DuPont, Cabot, Kemira, Millenium Chemicals, Degussa, Tioxide, Ishihara, Kerr-McGee, Coming Glass, General Electric, Lucent (former AT and T Bell Labs) etc.

In 2004, the liquid and gas-based processes to manufacture fine particles on an industrial level cover an estimated world wide market volume of at least 15 billion US dollars per year. With the emergence of numerous new products and nanoparticulate materials, it has been repeatedly speculated that dry processes are more economic and environmentally friendlier than their wet counterparts due to fewer process steps. A study on process energy consumption and environmental loads have shown that the total energy consumption of the sulfate process for TiO_2 production is about 1.68~2.1 times higher than that of the chloride process; and CO_2 emissions of the sulfate process is about 1.25 times higher than that of the chloride process as given in Table 1. Moreover, aerosol processes have the advantages in high purity, high yield, easy separation of the product powder from the exhaust, and have the potential advantages to provide rapid mixing on a molecular scale for nanocomposite materials.

The control of particle characteristics during flame synthesis is crucial because the properties of materials made from these particles depend on size and size distribution, morphology, extent of aggregation and chemical and phase composition. For example, in the manufacture of titania pigments, the goal is to produce a nearly monodisperse rutile phase particle about 250 nm in diameter resulting in maximum hiding power per unit mass. In contrast, in manufacture of powders for structural ceramics, particle size may not be so important though the size distribution may affect powder flow and packing prior to compaction. There, aggregates should be avoided since they result in pores and flaws during sintering and, thus, reduce the strength of the final part or specimen made with these particles.

Although this is an established industrial process bringing sizable profits to the corresponding corporations, its fundamentals are not yet well understood. The reason for this stems from the fact that chemical reaction and particle growth take place extremely fast (under a second) in typical process conditions. These short residence times coupled with high process temperatures make it rather difficult to collect representative samples for particle characterization and model development. This lack of understanding makes the process development and scale-up for manufacture of titania, silica and other ceramic powders rather difficult. This is important since flame technology has high potential for inexpensive manufacture of ceramics, especially nanoparticles, that can be used for synthesis of a wide spectrum of new materials. Flame technology produces the cheapest submicron powder, titania, in the market today ($l/lb, including the finishing cost).

European Science Foundation in 1994 approved a five-year program from January 1995 to promote research activities in nanostructured particles and materials, bridges between the aerosol community and the materials science community. The scientific thrust of the research is the synthesis of ceramic aerosols and films using gas-phase techniques, with the aim of generating single phase, or nanodispersed structural ceramic materials and electroceramics with new or improved properties. The research themes concentrate on fundamental research problems needed to be solved for optimal synthesis in the gas phase of nanoparticles and nanostructured materials with tailored microstructures.

Theme 1: *Nanoparticle Synthesis in the gas phase*

In order to achieve the desired properties of the nanostructured materials, it is necessary to generate appropriate properties in the primary particles formed in the gas phase. Important properties are particle size and shape, as well as chemical composition, phase and crystallinity. Flame aerosol process is investigated in more detail by developing theoretical models and performing experiments.

Theme 2: *Characterization in the gas phase*

As the properties of nanostructured materials are strongly dependent on the properties of the primary particles in the gas phase. Therefore, and also to understand the formation processes and or process control purposes, on line and *in situ* techniques are aimed to be develop to characterize the particles in the gas phase.

Theme 3: *Development of particle collection processes*

To form a nanostructured material from the particles formed in the gas phase, the nanoparticles need to be collected from the gas phase. Separation processes of cyclones and filter bags are investigated including theoretical modeling and experimental work, and this will promote the control as well as the optimization of the processes.

Theme 4: *Interfacing materials and aerosol science*

Special emphasis is put on developing communicative tools and definitions for interfacing various disciplines. Scientists from the different disciplines are brought together, share their experience and expertise, and adjust their terminology.

Theme 5: *Nanostructured materials and applications*

The main emphasis in this ESF program is on nanostructured materials for application in rechargeable batteries, HT ceramic fuel cells and solar cells. But the synthesis methods and tailored (micro) structures to be developed are by no means limited to the class of electroceramic and semiconducting materials, the knowledge to benefit other applications as well.

Theme 6: *Chemical and physical properties*

Energy dispersive X-ray fluorescence, secondary ion-mass spectroscopy, Rutherford backscattering, and X-ray photoelectron spectroscopy are pointed to be used to determine the compositions of the nanoparticles and the nanostructured materials. Transmission and scanning electron microscopy (TEM, SEM) and atomic force microscopy (AFM) are used to investigate the microstructure of the nanostructured materials. X-ray diffraction and extended X-ray absorption fine-structure measurements are used to determine the atomic structure of, possibly amorphous, nanoparticles and nanostructured materials.

This monograph summaries authors' research work in recent ten years at Dalian University of Technology and Shanghai Second Polytechnic University, under the financial supports of Material Science and Engineering fund of Dalian University of Technology, the Second Shanghai Key Discipline Construction Plan of Shanghai Municipal Education Commission (P1701), and an innovative research program of Shanghai Education Committee (07ZZ180). The book forces on authors' research work on the synthesis and characterization of TiO_2 nanopaticles by flame CVD method, particle dynamics for the growth of aerosol particles and numerical simulation of particle growth in flame CVD process by computational fluid dynamics (CFD), and the application of photodegradation of volatile organic compounds (VOC).

The authors are grateful for the financial supports of Material Science and Engineering fund of Dalian University of Technology, the Second Shanghai Key Discipline Construction

Plan of Shanghai Municipal Education Commission (P1701), and an innovative research program of Shanghai Education Committee (07ZZ180), and the contributions of our graduates ZHANG Wei, HAO Xiaomei, WANG Lixi, LI Yun, SUN Yaoyi, FANG Chuanyan, ZHANG Yaning, ZHANG Lingpin, RONG Jie and many others. Special thanks to Profs. NING Guiling and LI Zhiyi of Dalian University of Technology who led us into the field of nanotechnology, and Prof. XIE Maozhao of Dalian University of Technology for his supervision in the numerical simulation of flame CVD processes by CFD. Last but not least, the authors would be much appreciative for Prof. YANG Haifeng of Shanhai Normal University for his helps and discussions in the characterization of TiO_2 and SnO_2 nanoparticles, and Ms. XU Qiaolian and Mr. ZHANG Liming of Dalian University of Technology for their helps in experiments, and Drs. CHEN Shenwen, ZHU Luping, WANG Lingling, YANG Dandan, MA Changwen, WANG Jingrong, GAO Guilan, BING Naici, YUAN Hao, TIAN Zhen and WANG Lijun of Shanghai Second Polytechnic University for their helps in experiments and careful checks for the monograph.

Chapter 1

SYNTHESIS OF NANOPARTICLES BY PROPANE/AIR FLAME CVD PROCESS

NOMENCLATURE

d_{50}	particle median size, m
d_{16}	particle size corresponding to 16% value, m
d_{84}	particle size corresponding to 84% value, m
D	burner diameter, m
I_A	relative intensity of anatase <101>
I_R	relative intensity of rutile <110>
k	Boltzmann constant, $1.3805*10^{-23}JK^{-1}$
K_c	constant defined by *eq.* 1-6
L	reactor or flame length, m
M_P	molecular weight of particles, kg $kmol^{-1}$
M_{pre}	molecular weight of precursors, kg $kmol^{-1}$
n_0	initial particle number density, m^{-3}
N_{grw}	particle growth number
T	temperature, K
u	macroscopic stream velocity, ms^{-1}
$\dot{V}$	total gas volumetric flow rate, m^3s^{-1}
x_R	rutile mass fraction

Greek Letters

λ	mean free path of gas molecule, m
σ	relative spread of size distribution

1.1. INTRODUCTION

Flame CVD method for nanoparticles, in general, comprises four parts: a reactor, an air compressor and dryers, precursor bubblers, and particle collection devises as shown by Figure 1-1. Fuel and oxidants are burnt by a burner inside the rector or combustion chamber, generating high temperature typically between 800~1200°C. Flame CVD method has been subjected intensive investigations on the generation of ceramic nanoparticles using different fuels of hydrogen or methane [1]. In this book, the results obtained by industry propane/air diffusion flames are presented based on authors' work in recent ten years.

Air is used as oxidant and carrier air for precursors and is compressed by an air compressor, regulated by a pressure regulator, and dried by a refrigerated dryer for burner and further dried by a silicon gel bed for carrier air.

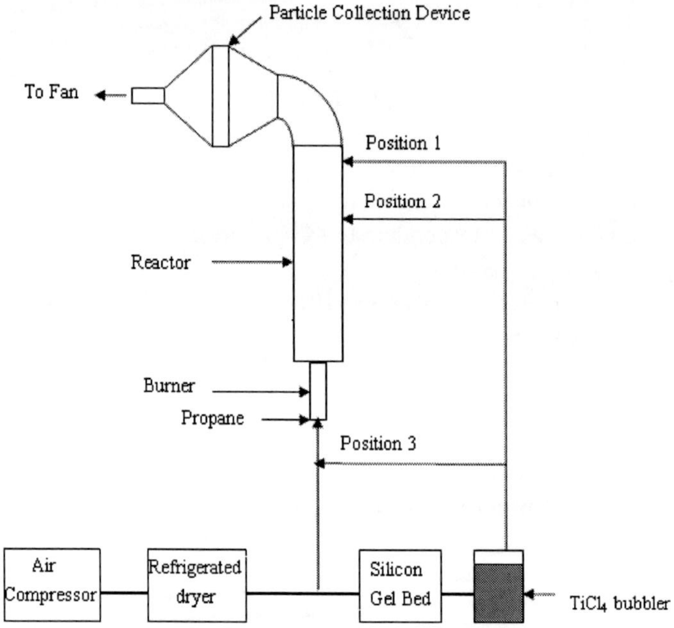

Figure 1.1. Schematic diagram of flame CVD experimental apparatus.

Precursors are carried into the reactor by carrier air and oxidation reactions occur inside the reactor leading to target nanoparticles. Precursors could be carried into the reactor at different positions as shown in Figure 1-1 in order to control the residence time of resultant nanopaticles in the reactor, and hence on particle properties such as size and size distribution, phase composition, UV adsorption and PL spectra and etc..

Particles are collected by a particle collector, which is just over the top end of the reactor. A 200 mesh stainless steel sicve is mounted inside the collector to collect nanoparticles for analysis and application studies. The exhaust gas is ventilated by an exhaust fan. Temperature profiles in the reactor are measured by thermocouples through tapings at different elevated levels of the reactor.

In experiments, target flow rates of air and propane are set by mass flow controllers and ignited to set up a flame heating up the experimental equipment. The carrier air for precursor is then set to a target value by a mass flow controller to dry the conveying pipeline for precursors, while precursor is not put into the bubbler. After the temperature attained 373K behind the particle collector, precursor is poured into the bubbler while the carrier air is stopped. After precursor is poured into the bubbler, the carrier air is restarted and the exhaust fan is switched on. The pressure of the reactor increases as nanoparticles deposited on the fine sieve increases. The exhaust fan is switched off when the pressure of the reactor is closed to atmospheric pressure (at about -2 ~ 5mmH_2O gauge pressure), the stainless steel sieve in the particle collector is taken out and the nanoparticles deposited on the sieve are collected. After each experimental run, the flow rate of the precursor is calibrated.

1.2. Carbon Doped TiO_2 Nanoparticles by Laminar Diffusion Flames

Laminar diffusion flames have the advantages of its simplicity and easy to perform experiments, and are frequently used by different authors to study flame CVD method for synthesis of nanoparticles. Figure 1-2 is an illustration of a burner of propane/air laminar diffusion flame for the synthesis of TiO_2 nanoparticles. The burner is made up of four concentric stainless-steel tubes (8×1, 13×1.5, 20×2, and 27×3mm in their nominal size and wall thickness).

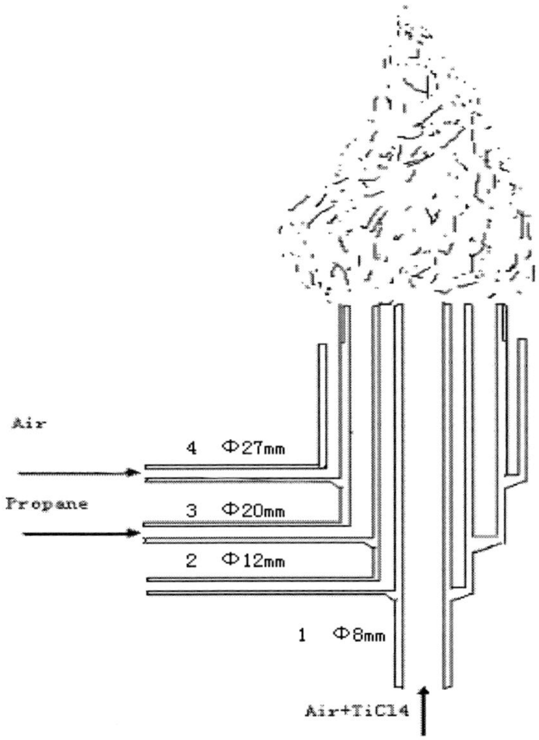

Figure 1.2. Illustration of a burner of propane/air laminar diffusion flame for the synthesis of TiO_2 nanoparticles.

Table 1.1. Experimental conditions used in the experiments

Run No.	Pressure (atm)	Flow Rate(mmol/s)			$TiCl_4$ Flow Rate (g/min)	Particle number density (m^{-3})
		C_3H_8	Air (oxidant)	Air (carrier gas)		
1	0.98	0. 129	5.539	0.749	0.0686	4.07E+20
2	0.98	0. 259	5.988	1.123	0.113	4.97E+20
3	0.98	0. 259	5.988	1.497	0.135	6.19E+20
4	0.98	0. 259	6.737	0.374	0.0703	3.0E+20
5	0.98	0. 259	8.234	0.374	0.0965	3.87E+20
6	0.98	0. 259	9.880	0.374	0.0965	3.4E+20
7	0.98	0. 259	11.078	0.374	0.079	2.55E+20
8	0.98	0. 259	11.078	1.123	0.113	4.17E+20
9	0.98	0. 259	10.78	1.497	0.135	5.24E+20
10	0.98	0. 259	11.976	0.374	0.079	2.43E+20
11	0.98	0. 388	4.491	0.374	0.07	3.26E+20
12	0.98	0. 388	5.614	0.374	0.06	2.36E+20
13	0.98	0. 388	6.737	0.374	0.06	2.0E+20
14	0.98	0. 388	7.859	0.374	0.06	1.82E+20
15	0.98	0. 388	8.982	0.374	0.06	1.63E+20
16	0.98	0. 388	8.982	0.749	0.076	2.09E+20

Industrial propane is used as fuel and is fed into the third tube from the central tube; air is used as oxidant and is fed into the fourth tube to generate an industrial propane/air diffusion flame. TiO_2 nanoparticles are generated in the industrial propane/air laminar diffusion flame by the oxidation of $TiCl_4$, which is the precursor of TiO_2 nanoparticles and is carried into the central (first) tube of the burner by dry air through a bubbler. The fourth tube is 0.05m lower than the others and the outer side of the fourth tube is screwed to stabilize the flame. The combustion chamber is a 0.80m long, 157×3.5mm stainless steel pipe.

A Pt-Rh R type thermocouple (Omega Engineering) is arranged about 3 cm above the top of the burner to measure the flame temperature. Sixteen experimental runs at different operational conditions are carried out and the operational conditions are tabulated in Table 1-1. The mole ratio of oxygen to propane is ranging from about 2.5 to 9.0, and the mole fraction of precursor $TiCl_4$ in the flame is ranging from about 0.0005 to 0.0016.

1.2.1. Flame Properties

For laminar diffusion flames, the flame length in general, is proportional to fuel flow rate and inverse proportional to gas diffusion coefficient, and is affected by the stoichiometric ratio and oxygen fraction in the flame. The flame length decreases as oxygen fraction in the flame and stoichiometric ratio increase.

Effects of the mole ratio of O_2 to C_3H_8 on visual flame length and flame temperature are shown in Figures 1-3 and 1-4, respectively. The stoichiometric mole ratio of O_2 to C_3H_8 is 5.

It can be seen that the flame length increases from about 0.04m to 0.2m corresponding the mole ratio O_2 to C_3H_8 decreases from about 6 to about 2.5 due to incomplete combustion, but flame length changes little when the mole ratio of O_2 to C_3H_8 is above 6. The flame temperature decreases from about 1750 K to 1400 K as the mole ratio of O_2 to C_3H_8 increases from about 2.5 to 6 and then decreases to about 1000K when the mole ratio of O_2 to C_3H_8 increases up to 9.

1.2.2. Particle Morphology

1.2.2.1. Mean Particle Size

The median size (d_{50}), obtained from the graph of cumulative fraction undersize versus size, of the 16 test runs is ranging from about 20-60nm. Size distributions of number frequency and cumulative fraction undersize versus size are obtained by counting up particles from TEM (TEC NAI G^2 20) pictures. 100-300 particles are counted up for each test run. Typical TEM pictures are given in Figure 1-5 to show the effects of the mole ratio of propane to air on particle morphology. It can be seen that the particles are spherical at high temperature of about 1695K, but are in regular shape at low temperature of about 1005K. Experimental results show that particles are spherical at flame temperatures above 1250K, and irregular shaped particles occur at flame temperatures below 1250K and more irregular shaped particles are formed at lower flame temperatures.

The mean particle size and size distributions at different operation conditions are determined by counting more than 100-300 particles from their TEM pictures. The mean particle size is in the range from about 20nm to 60nm under the experimental conditions in Table 1. The mean particle size versus particle growth number N_{grw} is plotted in Figure 1-6.

Literature data for other nanoparticles from different flames is also presented in the figure. It could be seen that the mean primary particle size grossly correlated well with the particle growth number as

$$d_{50} = 1.35(1 + N_{grw})^{1/3} \tag{1-1}$$

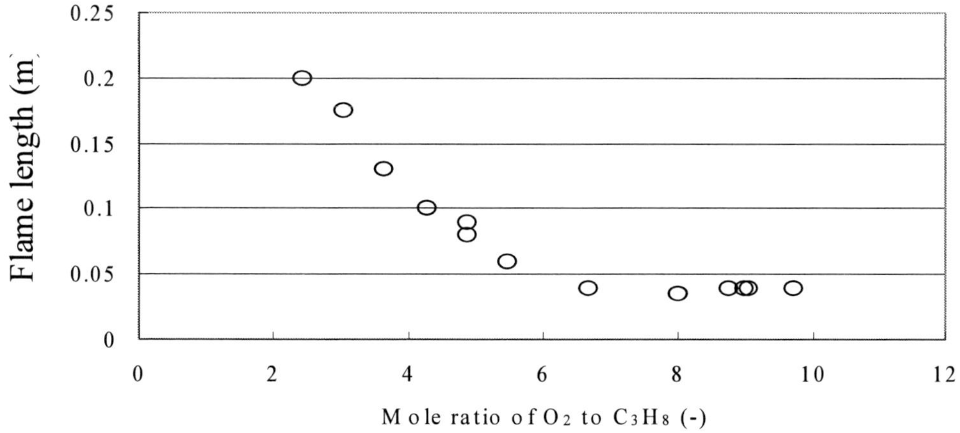

Figure 1.3. Flame length *versus* O_2/C_3H_8 mole ratio.

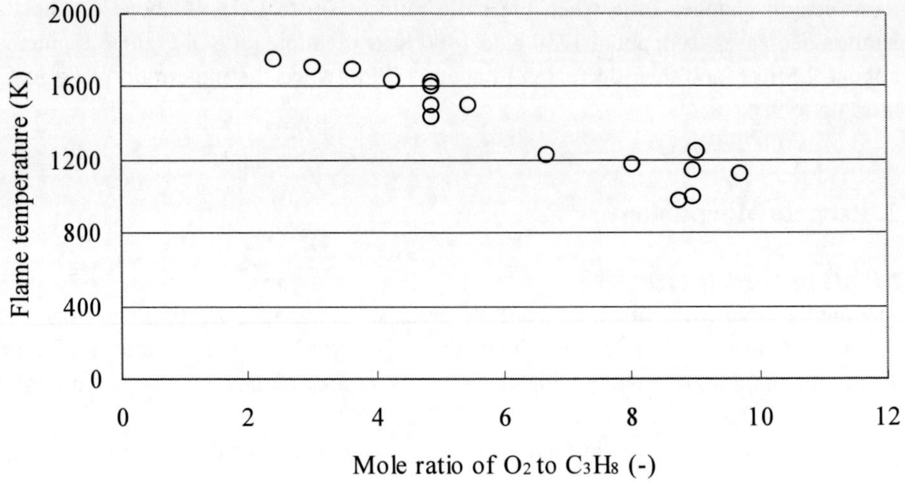

Figure 1.4. Flame temperature *versus* O_2/C_3H_8 mole ratio.

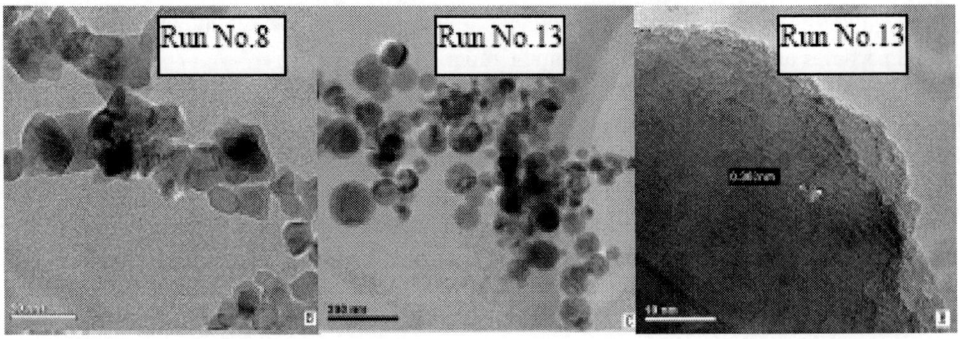

Figure 1.5. Typical TEM images of TiO_2 nanoparticles.

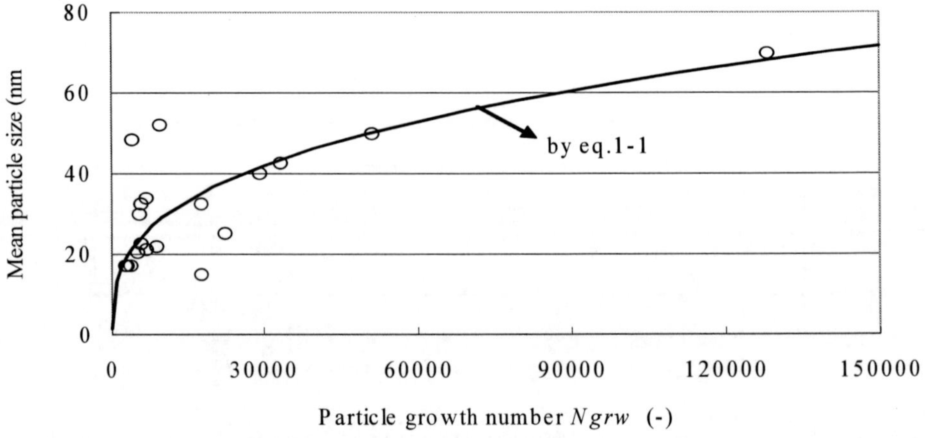

Figure 1.6. Mean particle size *versus* particle growth number N_{grw}.

The particle growth number is

$$N_{grw} = n_0 \sqrt{\frac{\lambda kT}{\rho_P}\frac{L}{u}} \tag{1-2}$$

where λ is mean free path of gas molecules, k is the Boltzman constant, T is flame temperature, ρ_P is particle density, L is the flame length, u is propane velocity at the burner outlet at the flame temperature, and n_0 is the particle number density at the burner outlet:

$$n_0 = \frac{6}{\pi} \times 10^{27} \frac{M_{pre}}{M_P} \frac{\dot{W}_{pre}}{\dot{V}\rho_P} \tag{1-3}$$

where M_{pre} and M_P are the molecular weights of precursor $TiCl_4$ and product TiO_2, respectively, $\dot{V}$ is the total gas volume flow rate, and $\dot{w}_{pre}$ is the mass flow rate of the precursor $TiCl_4$. The particle number density at the burner outlet n_0 for the 16 test runs are also given in Table 1 and it is a vital parameter to control the mean particle size. From above equations, mean particle size depends largely on the mass flow of the precursor $TiCl_4$, $\dot{w}_{pre}$, or on the mole fraction of $TiCl_4$ in the flame.

The average gas velocity inside flame u is equal to

$$u = \frac{\dot{V}}{\frac{\pi}{4}D^2} \tag{1-4}$$

where D is burner diameter and is 27mm for present experiments.

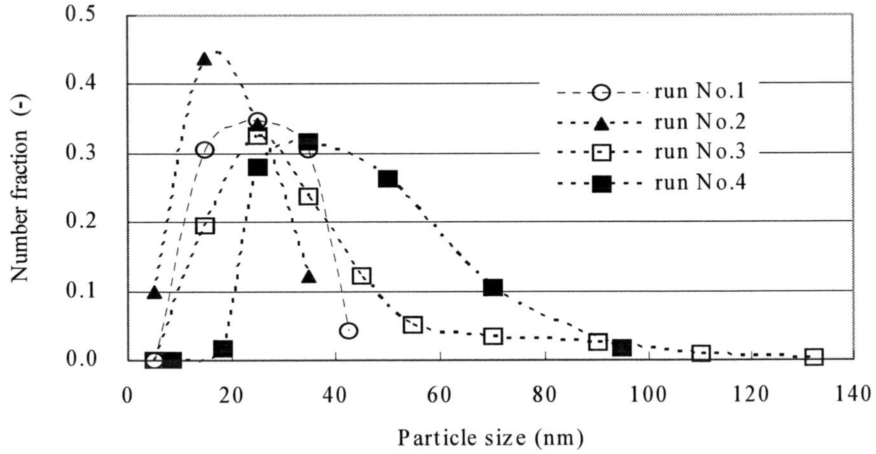

Figure 1.7. Number frequency size distribution of runs No.1-4.

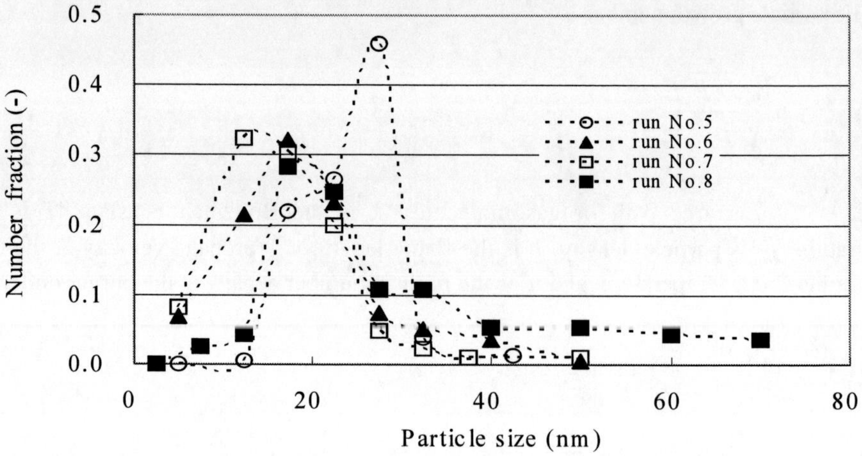

Figure 1.8. Number frequency size distribution of runs No.5-8.

1.2.2.2. Size Distributions

Size distributions of number frequency and cumulative fraction undersize versus size of the 16 test runs are shown in Figures 1-7～1-10 and Figures 1-11～1-14, respectively.

It can be seen that the size distribution is close to the normal distribution, and relative spread of size distribution is between 0.2-0.6. The width of the size distribution is fairly wide based on the relative spread [2]. The relative spread of size distribution σ is given by [2]

$$\sigma = \frac{d_{84} - d_{16}}{2d_{50}} \tag{1-5}$$

where $d_{84\%}$ and $d_{16\%}$ are the particle sizes corresponding to 84% and 16% values, respectively, on the graph of cumulative fraction undersize versus size.

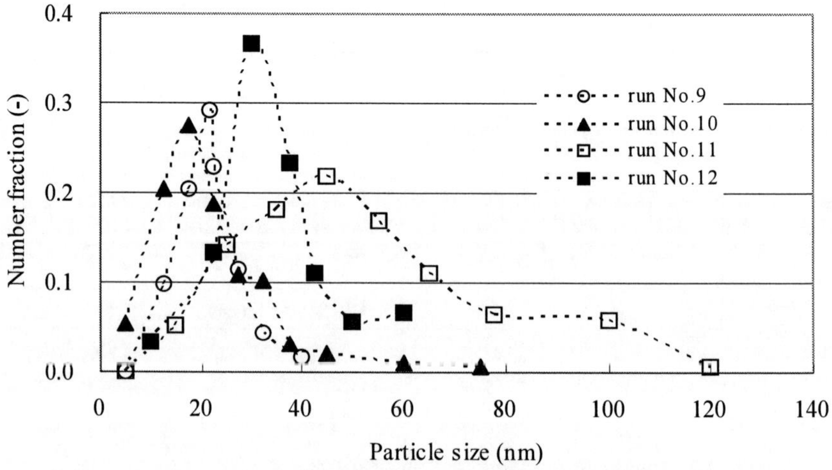

Figure 1.9. Number frequency size distribution of runs No.9-12.

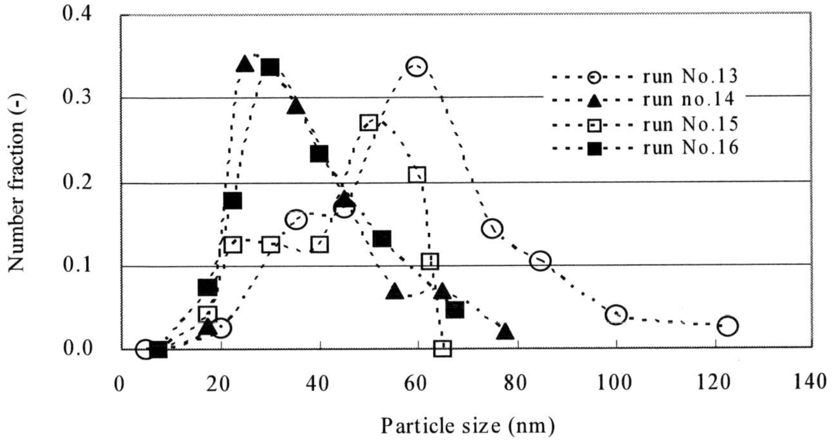

Figure 1.10. Number frequency size distribution of runs No.13-16.

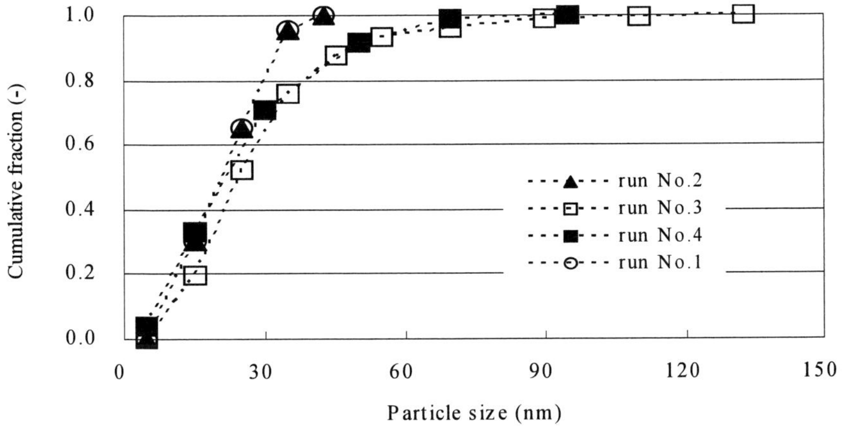

Figure 1.11. Cumulative size distribution of runs No.1-4.

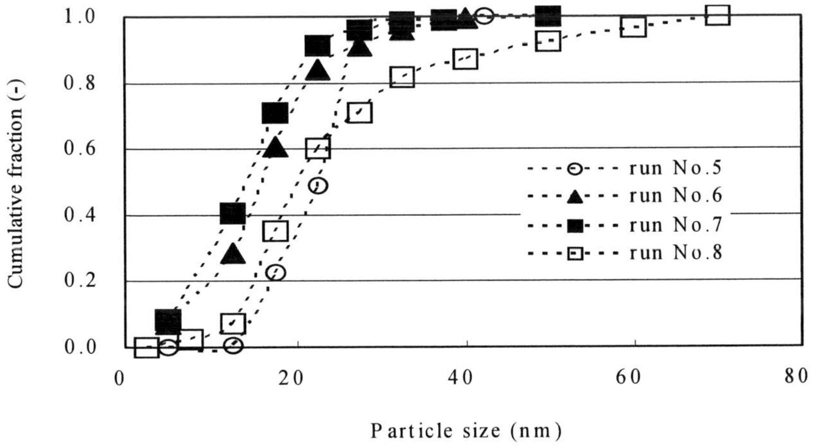

Figure 1.12. Cumulative size distribution of runs No.5-8.

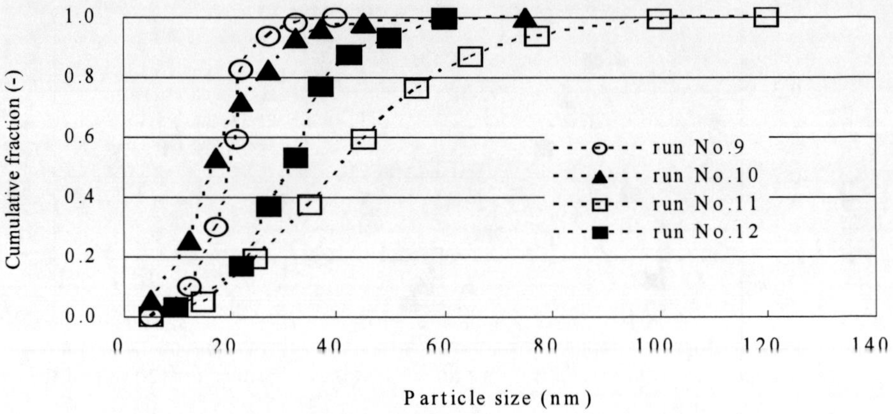

Figure 1.13. Cumulative size distribution of runs No.9-12.

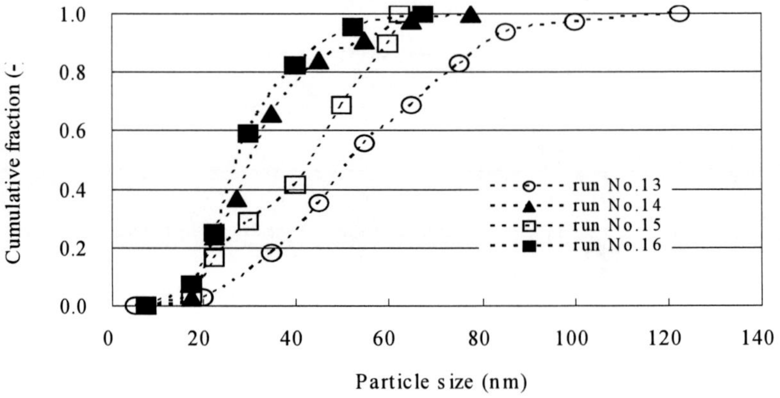

Figure 1.14. Cumulative size distribution of runs No.13-16.

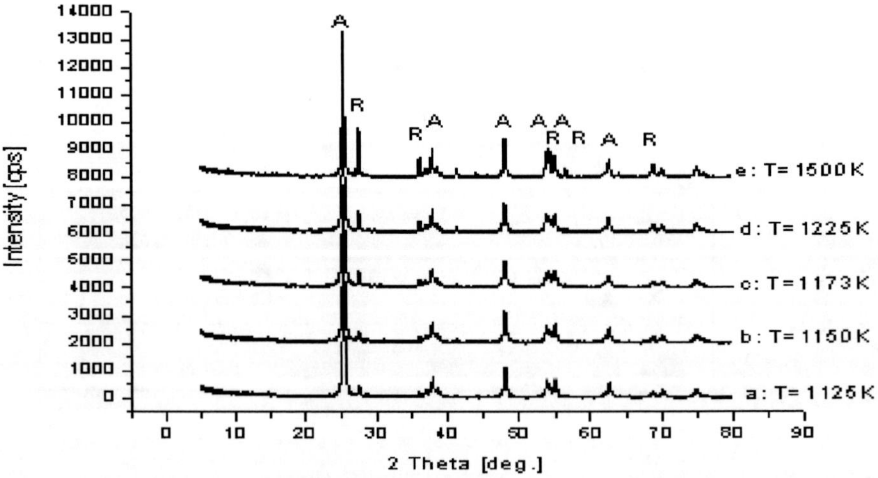

Figure 1.15. XRD patterns at different temperatures.

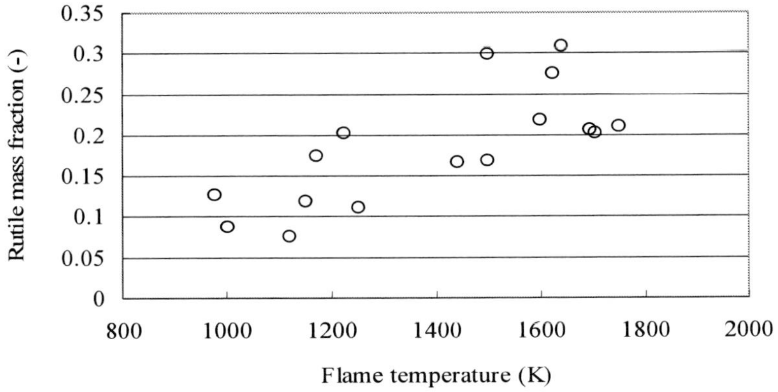

Figure 1.16. Rutile mass fraction *versus* flame temperature.

1.2.3. Phase Composition

The rutile mass fraction is obtained from XRD patterns, which are measured by a Rigaku D-MAX/A diffractometer. Typical XRD patterns are shown in Figure 1-15. The figure shows that the particles consists primarily anatase phase, and the intensity of rutile phase raises as temperature increases, showing transformation of anatase to rutile as increasing reaction temperature provides the activation energy for transformation from anatase phase to rutile phase [3]. The mass fraction of rutile phase versus flame temperature is plotted in Figure 1-16 and it shows that the mass fraction of rutile phase depends mainly on flame temperature and increases from about 0.07 to 0.3 corresponding to the flame temperature from about 1000K to 1700K. The mass fraction of rutile phase is estimated by [4]

$$x_R = \frac{1}{1 + K_c I_A / I_R} \qquad (1\text{-}6)$$

where I_A and I_R are the relative intensity corresponding to anatase <101> and rutile <110> reflection, respectively, and K_c is a constant of 1.26.

1.2.4. Carbon Dots

Because of the incomplete combustion of the industrial propane in air, carbon may form in the process so that the titania nanoparticles could be mixed with carbon. It has been proven that carbon mounted or mixed titania nanoparticles have higher performance in photocatalytic reactions than their unmixed counterparts because of the high adsorption ability of carbon nanoparticles [5,6]. Carbon content in the product is determined by a Shimadzu EPMA1600 electron probe microanalyzer, typical EPMA patterns are shown in Figure 1-17. The carbon content obtained by EPMA relative intensity is given in Table 1-2 and is ranging from zero to about 20% by weight under different operational conditions.

Experimental and theoretical studies have shown that more carbon is formed at low oxygen to fuel ratio for hydrocarbons due to incomplete combustion, and the formation of carbon is related to acetylene concentration in flame [7,8].

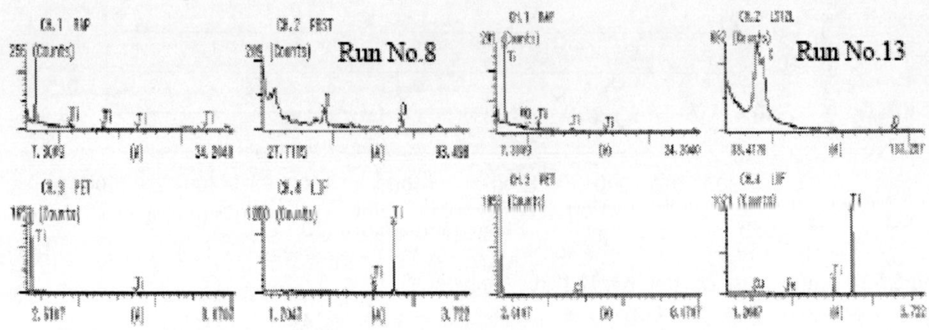

Figure 1.17. Typical EPMA patterns at different experimental conditions.

However, present results show that the carbon formation depends not only on the mole ratio of O_2 to C_3H_8, but also depends on the particle number density n_0 as can be seen from Figure 1-18, which may be due to the cause of that TiO_2 particles may lead to the decomposition of acetylene. Therefore, carbon content could be controlled by the combination of the mole ratio of O_2 to C_3H_8 and the particle number density n_0.

1.2.5. UV Absorptivity

Typical UV absorption spectra are shown in Figure 1-19. The measurements are carried out at a concentration of TiO_2 aqueous solution of 2.5mmol/L. The absorptivity is reached to almost constant value when the wavelength is down to smaller than 320nm.

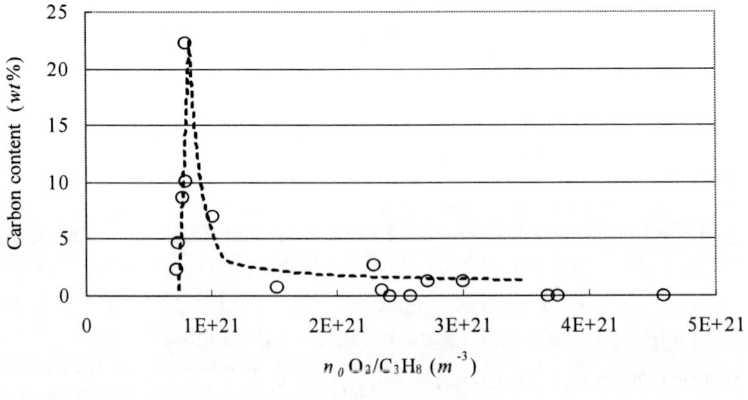

Figure 1.18. Carbon content *versus* particle number density times the mole ratio of O_2 to C_3H_8.

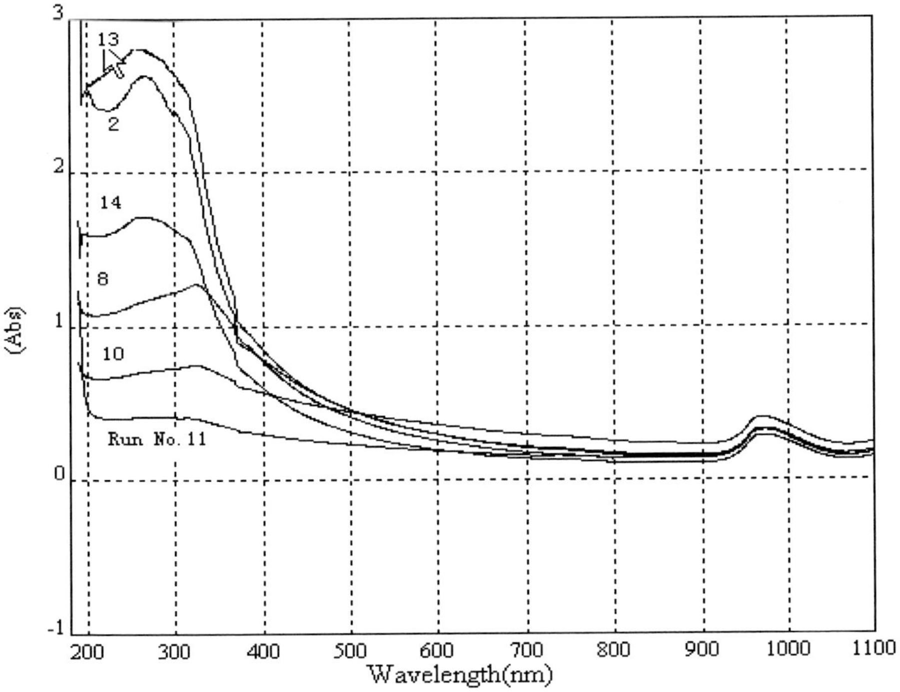

Figure 1.19. Typical UV adsorption spectra.

Table 1.2. Experimental results of laminar diffusion flame

Run No.	Flame length (m)	Flame temperature (K)	Mean particle size (nm)	Rutile mass fraction (-)	Carbon content % (wt)	UV adsorptivity TiO_2: 2.5mmol/L (Abs)
1	0.04	1250	24.3	0.072	0.0	1.163
2	0.08	1500	16.6	0.112	0.0	2.528
3	0.08	1440	32.3	0.111	1.324	1.190
4	0.06	1500	42.9	0.210	0.828	2.000
5	0.04	1225	23.6	0.136	0.0	1.053
6	0.035	1173	23.5	0.117	1.358	1.244
7	0.04	1150	16.6	0.078	2.719	1.680
8	0.04	1005	24.8	0.056	0.0	1.272
9	0.04	980	26.0	0.083	0.0	1.182
10	0.04	1120	16.5	0.049	0.459	0.745
11	0.20	1753	61.9	0.143	22.29	0.406
12	0.175	1708	56.7	0.136	2.28	1.319
13	0.13	1695	56.5	0.139	4.618	2.693
14	0.10	1640	48.9	0.219	8.665	1.696
15	0.09	1623	37.7	0.192	10.083	1.081
16	0.09	1600	39.3	0.148	7.023	1.215

The averaged peak values of absorptivity are given in Table 1-2 for the particles synthesized at different operational conditions. Effects of rutile mass fraction on the averaged peak values of absorptivity are shown in Figure 1-20, and the figure shows that the averaged peak values of absorptivity increases with increasing rutile mass fraction. Effects of carbon content on the averaged peak values of absorptivity are shown in Figure 1-21, it could be seen that the averaged peak values of absorptivity occurres a maximum at carbon content of about 5% by weight (sample No.13).

Experimental results of photodegradation of gas pollutants show that the degradation degree increases with increasing rutile mass fraction and occurred a maximum at carbon content of about 5% by weigh, which will be shown in Chapter 4, coincides with the effects of rutile mass fraction and carbon content on UV absorptivity. It is, therefore, the mole ratio of O_2 to C_3H_8 and the particle number density n_0, or the mass flow of the precursor $\dot{w}_{TiCl_4}$, or the mole fraction of $TiCl_4$ in the flame are vital parameters to control rutile mass fraction, carbon content, as well as mean particle size of the carbon doped titania nanoparticles.

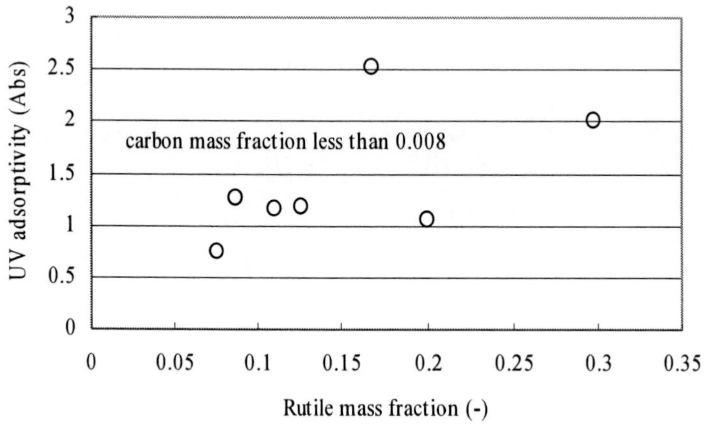

Figure 1.20. UV adsorptivity *versus* rutile mass fraction.

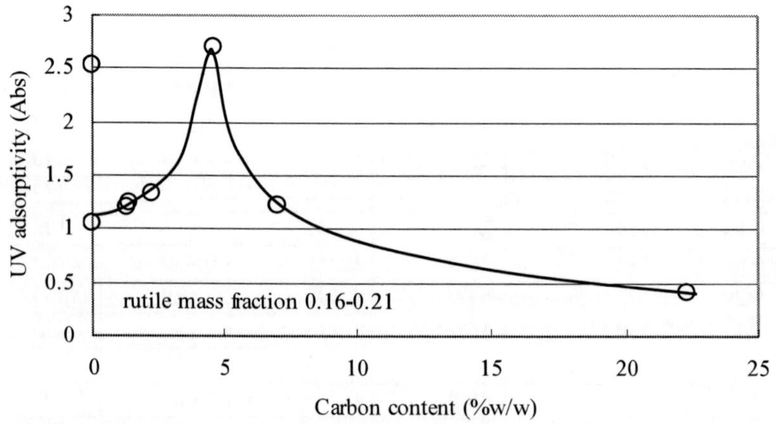

Figure 1.21. UV adsorptivity *versus* carbon content.

1.3. TiO_2 NANOPARTICLES BY TURBULENT DIFFUSION FLAME

Laminar diffusion flames have the advantages of its simplicity and easy to perform experiments, but laminar flames have low combustion strength and thus it is difficult to scale up for mass production of nanoparticles by laminar diffusion flame CVD processes. Since turbulent diffusion flames have higher combustion strength, turbulent diffusion flame CVD process has potential abilities for mass production of nanoparticles.

The burner for propane/air turbulent flame is a Honeywell commercial burner of 2 thousand calories of thermo-capacity. The reactor or combustion chamber is a 0.50m long, 157×3.5mm stainless steel pipe.

$TiCl_4$ is carried into the reactor at three different positions to examine the residence time of TiO_2 nanopaticles in the reactor on particle size, phase composition, UV absorption and PL spectra. Position 1 is that the taping is at the level of 50mm to the top end of the reactor, position 2 is that the taping is at the level of 192mm to the top end of the reactor, and position 3 is that $TiCl_4$ is carried into the reactor through the burner.

Temperature profiles in the reactor are measured by using a S-type thermocouple. 8 tapings at levels from the top end of the reactor of 43, 116, 197, 236, 275, 315, 336 and 360mm are arranged to measure the temperature profiles axially. In the measurement, temperatures have been measured at different radial positions of 0, 6, 12, 18, 24, 39, 54 and 69mm from the central axes. Particles are collected by a particle collection devise, which is just over the top end of the reactor.

Table 1.3. Operational conditions in the turbulent diffusion flame experiments#

Run No.	Flow Rate(mol/s)			Excess air ratio (-)	Reactor Heat Flux (kJ/m^2s)	$TiCl_4$ Flow Rate (g/min)	$n_0^{\$}$ $\times 10^{20}$ (m^{-3})	Position*
	C_3H_8	Air (oxidant)	Air (carrier gas)					
1	0. 00583	0.119	0.0372	1.072	626.9	2.6	5.710	1
2	0. 00531	0.119	0.0372	1.177	520.1	2.6	6.419	1
3	0. 00479	0.119	0.0372	1.304	423.2	2.6	6.745	1
4	0. 00497	0.104	0.0372	1.136	504.0	2.6	7.739	1
5	0. 00419	0.104	0.0372	1.348	358.2	2.6	8.566	1
6	0. 00419	0.0908	0.0372	1.222	395.1	2.6	8.708	1
7	0. 00419	0.0908	0.0521	1.364	353.9	3.2	9.915	1
8	0. 00350	0.0908	0.0372	1.436	275.7	2.6	9.713	1
9	0. 00350	0.0908	0.0521	1.633	247.0	3.2	10.020	1
10	0. 00479	0.119	0.0372	1.304	423.2	2.6	6.745	2
11	0. 00479	0.119	0.0223	1.180	467.8	1.075	3.011	2
12	0. 00479	0.119	0.0149	1.118	493.7	0.740	2.151	2
13	0. 00419	0.0908	0.0372	1.222	395.1	2.6	8.708	3

#: Experiments were performed under atmospheric pressure and ambient temperature of 289K.

*: Positions of $TiCl_4$ carried into the reactor.

$: initial TiO_2 particle number density.

A 200 mesh stainless steel sieve with a cross section of 300mm×300mm could be inserted inside the collection devise to collect TiO_2 nanopaticles. The exhaust gas is ventilated by an exhaust fan. Thirteen experimental runs at different operational conditions are carried out and the operational conditions are tabulated in Table 1-3. The excess air ratio, α, is in a range from about 1.07 to 1.6. Reactor heat flux, $LHV/A\alpha$, which is the heat input (LHV of the fuel propane) divided by reactor cross sectional area A and excess air ratio α, is ranging from 247 to 627 kJ/m^{2}s. And the flow rate of precursor $TiCl_4$ is ranging from about 0.74 to 3.2 g/min.

1.3.1. Temperature Profiles

Measured temperatures are plotted against radial position in Figure 1-22 at different height levels for test runs 1, 5 and 9.

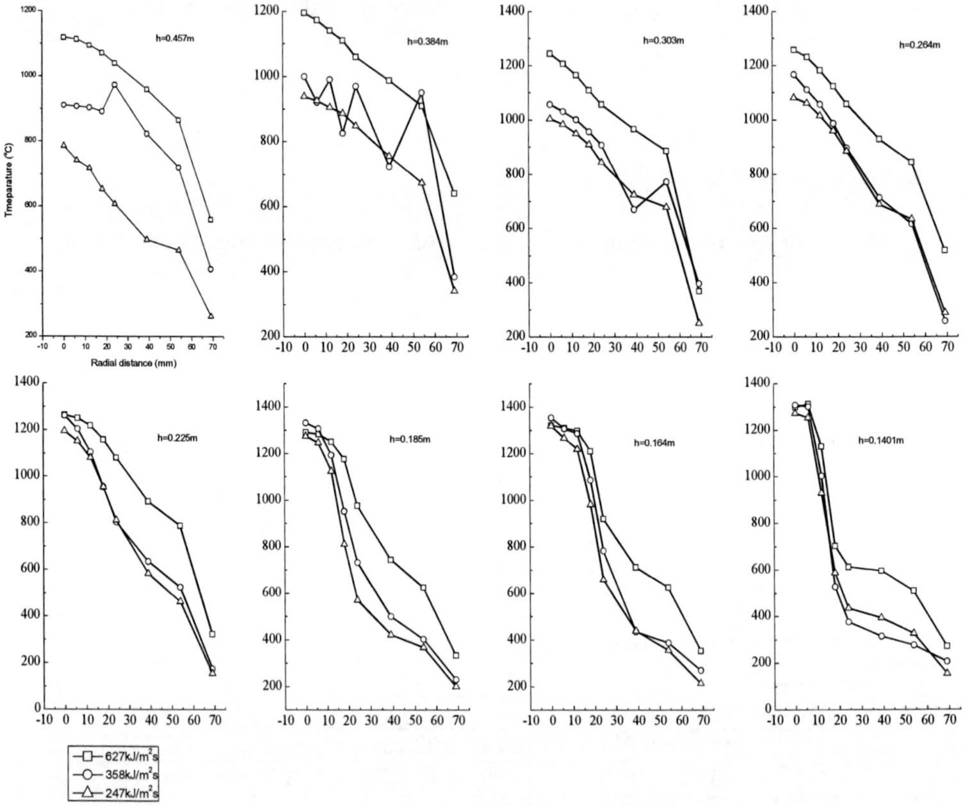

Figure 1.22. Measured temperatures *versus* radial position at different height levels.

The figure shows that temperature increases as reactor heat flux increases, decreases significantly towards the reactor wall, but becomes more uniform towards the top end of the reactor. Average temperatures in the core region and the entire reactor space, taken as the algebraic mean of entropy, are shown in Figure 1-23 against reactor heat flux. Figure 1-23 shows that the average temperature increases almost linearly with reactor heat flux. The column with diameter of 48mm is taken as the core region. The burner diameter is about 24mm.

The average temperature in the core region is ranging from 1046 to 1285K and the average temperature in the entire reactor space is ranging from 753 to 1067K, corresponding to reactor heat flux from 247 to 627 kJ/m^{2}s.

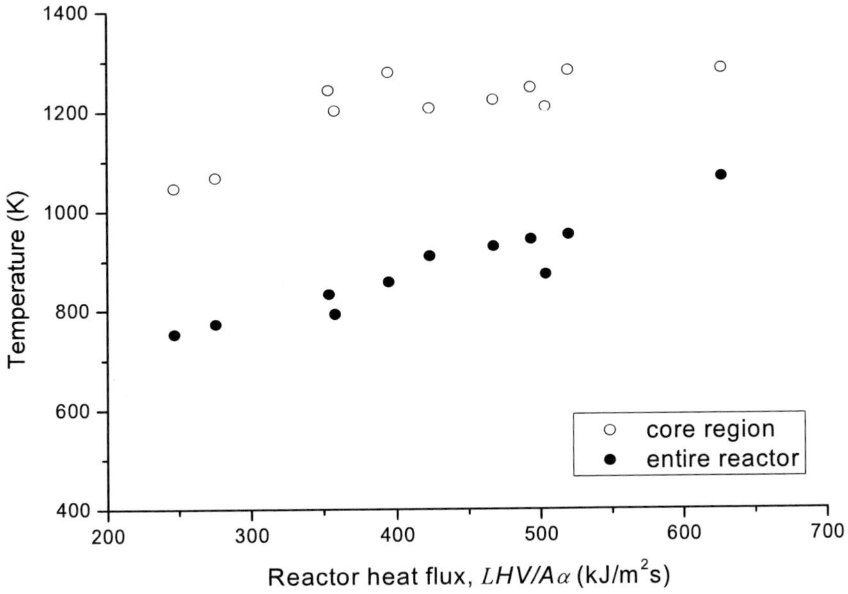

Figure 1.23. Average temperatures in core region and entire reactor space *versus* reactor heat flux.

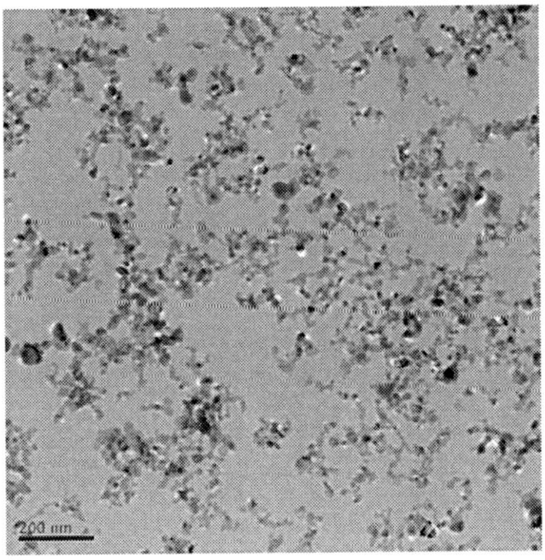

Figure 1.24. A typical TEM picture (run 4).

1.3.2. Particle Morphology

Spherical TiO_2 nanoparticles are produced under the operational conditions. A typical TEM picture (by FEI TECNAI G^2 S-TWIN) is given in Figure 1-24. A typical particle size distribution of number frequency and cumulative fraction undersize versus size is shown in Figure 1-25. It can be seen from the figure that the particle size distribution is close to the normal distribution, similar to the size distribution for TiO_2 nanoparticles synthesized by laminar diffusion flame CVD process. The median size (d_{50}) at different operational conditions is in the range from about 30 to 80nm as can be seen from Table 1-4. Mean particle sizes of test runs of 2 and 13 are obtained by BET method, the two mean sizes of run 2 obtained by the two different methods are in good agreement.

It can be seen from Table 1-4 that when $TiCl_4$ is carried into the reactor at Position 3, particle size is larger than that when $TiCl_4$ is carried into the reactor at Position 2. When $TiCl_4$ is carried into the reactor at Position 2, particle size is larger than that when $TiCl_4$ is carried into the reactor at Position 1. This is because of that particle growth due to particle coalescence increases as particle's residence time in reactor increases, which will be discussed in Chapter 2.

At about the same residence time, particle size increases as $TiCl_4$ flow rate increases, again due to particle coalescence mechanism, because TiO_2 particle number density increases as $TiCl_4$ flow rate increases and thus collisions between TiO_2 particles become stronger. Mean particle size versus particle growth number, N_{grw}, is plotted in Figure 1-26 including the data from TiO_2 nanoparticles synthesized by laminar diffusion flame CVD process. It could be seen that mean particle size grossly correlated well with the particle growth number, N_{grw}, by equation (1-1).

Table 1.4. Experimental results of turbulent diffusion flame

Run No.	T (K)	Apparent Residence Time# (s)	UV Adsorptivity (Abs/nm)	Particle Size (nm)	Rutile Mass Fraction (-)
1	1067	0.0579	1.374/270	47.5	0.487
2	953	0.0651	1.412/261	44.4/39.1*	0.427
3	909	0.0684	1.407/260	37.6	0.424
4	873	0.0785	1.163/296	42.0	0.401
5	793	0.0868	1.168/292	31.6	0.345
6	858	0.0883	1.159/295	30.5	0.363
7	833	0.0817	1.159/295	33.8	0.259
8	773	0.0985	1.142/300	31.6	0.231
9	753	0.0908	-	37.0	0.155
10	909	0.263	1.398/267	76.3	0.424
11	929	0.283	1.341/262	60.4	0.523
12	943	0.294	1.307/265	46.4	0.551
13	858	0.883	1.098/287	75.9*	0.575

$: Average temperatures taken as algebraic mean of entropy in reactor.

#: Apparent residence time of TiO_2 in reactor.

*: Obtained by BET method.

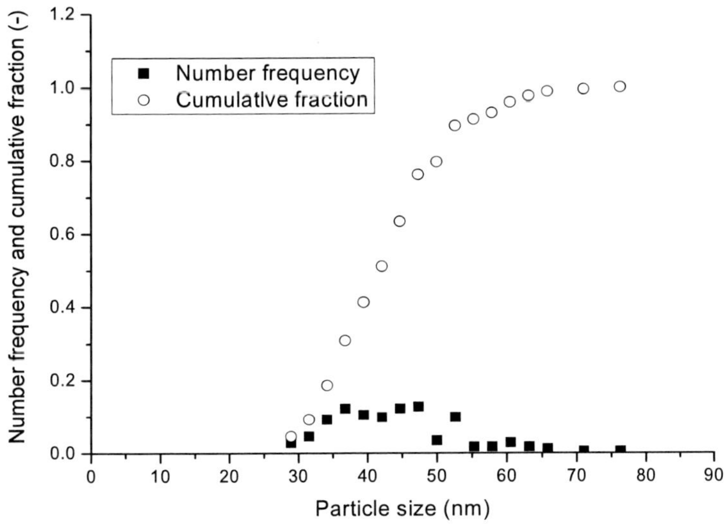

Figure 1.25. Particle size distribution of number frequency and cumulative fraction (run 4).

1.3.3. Phase Composition

Typical XRD patterns are shown in Figure 1-27. The figure shows that the intensity of rutile phase rises as reactor heat flux increases. 0.155 to 0.575 of rutile mass fraction is obtained corresponding to reactor heat flux from 247 to 627 kJ/m^{2}s.

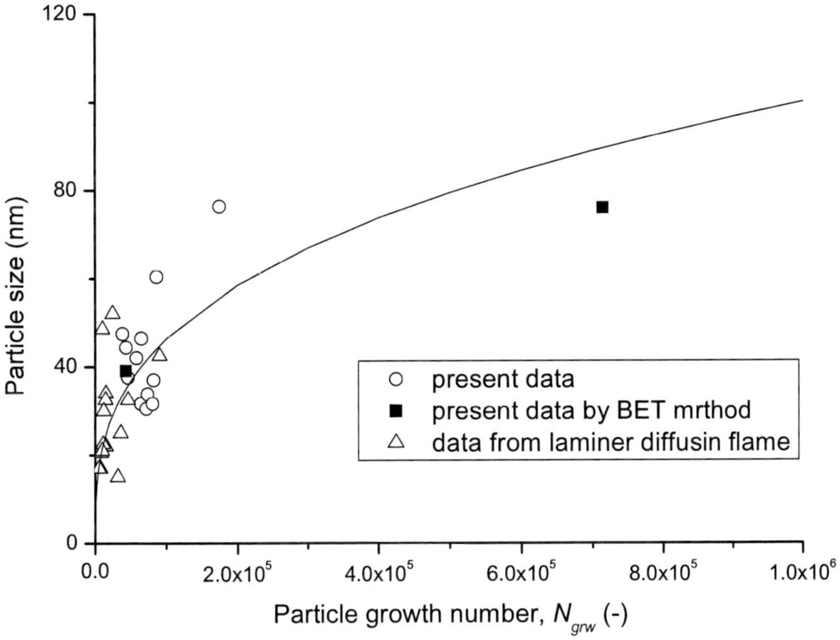

Figure 1.26. Mean particle size *versus* particle growth number, N_{grw}.

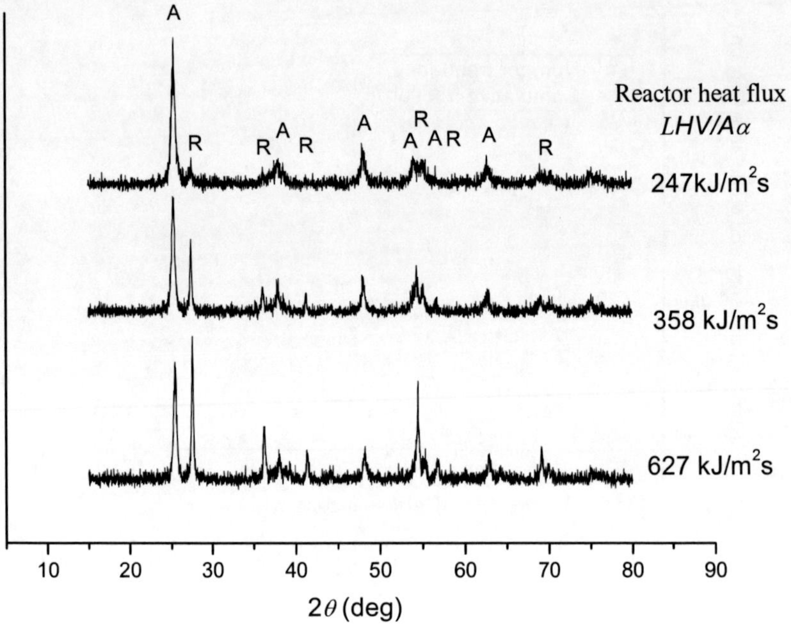

Figure 1.27. Typical XRD patterns.

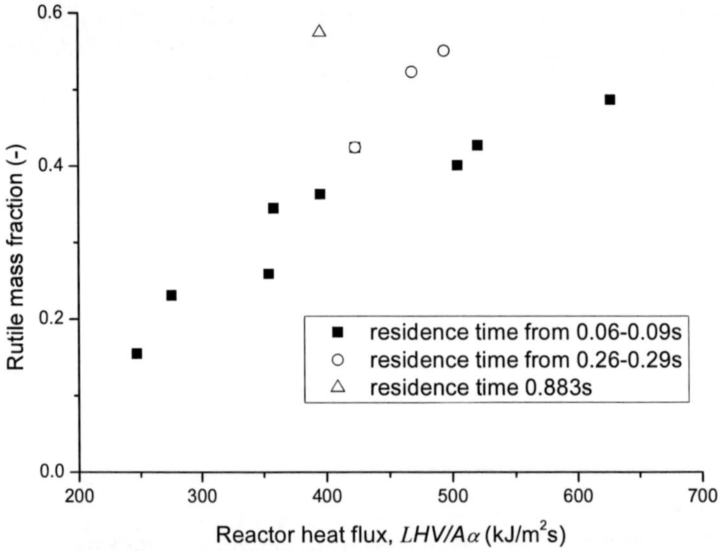

Figure 1.28. Rutile mass fraction *versus* reactor heat flux.

The mass fraction of rutile phase versus reactor heat flux is plotted in Figure 1-28 and the figure shows that the mass fraction of rutile phase depends not only on reactor heat flux, but also on the residence time of TiO_2 in the reactor, showing transformation of anatase to rutile as increasing reaction temperature provides the activation energy for transformation from anatase to rutile phase, and increasing residence time provides longer transformation process [3,9].

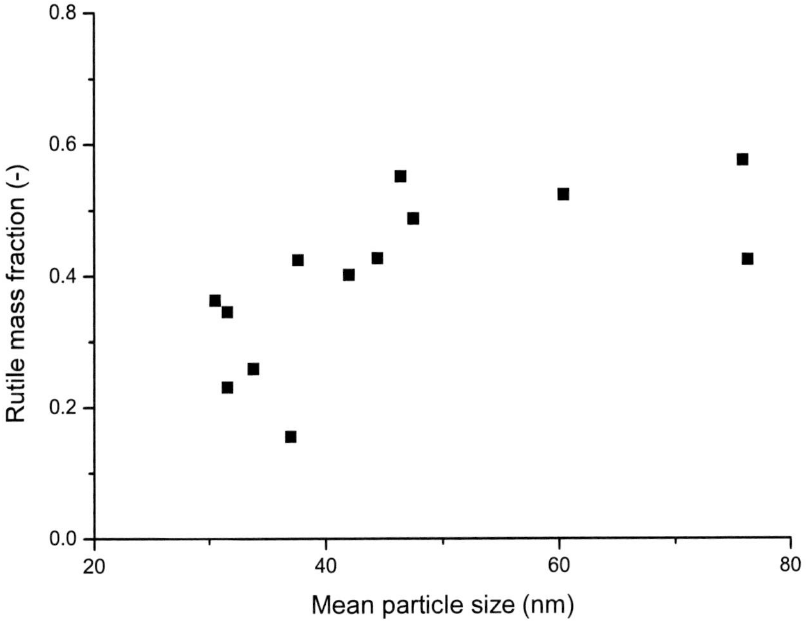

Figure 1.29. Rutile mass fraction *versus* particles size.

Rutile mass fraction versus particles size is shown in Figure 1-29 and it could be seen that rutile mass fraction grossly increases with increasing particle size, implying that anatase to rutile phase transforms as particle coalesces and growth.

1.3.4. UV Absorptivity

Typical UV absorption spectra are shown in Figure 1-30. The absorptivity starts to increase when wavelength is down to about 400nm and is reached to almost constant value when the wavelength is further down to smaller than 320nm. The peak values of absorptivity and corresponding wavelengths are given in Table 1-4. The peak values of absorptivity are from 1.1 to 1.4, and the corresponding wavelengths are from 260 to 300nm. All the TiO_2 nanoparticles under the thirteen experimental runs show typical TiO_2 nanoparticles' ultraviolet absorption spectrum.

1.3.5. PL Spectra

PL spectrum is an effective way to investigate the electronic structure and optic characteristics of semiconductor nano-materials, by which the information such as surface oxygen vacancies and defects as well as separation and recombination of photoinduced charge carries can be obtained [10]. Results of photodegradation of phenol by TiO_2 nanoparticles have shown that the photocatalytic activity is strongly related to the intensity of PL spectra [10].

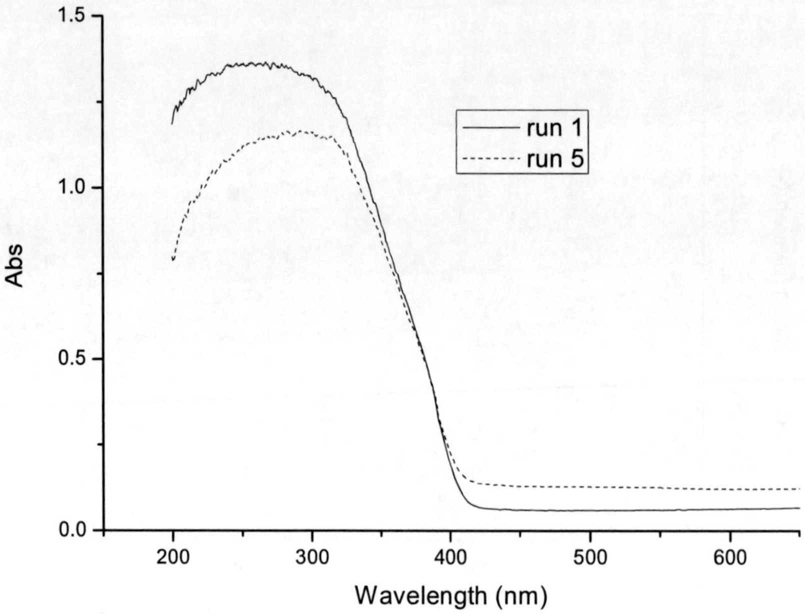

Figure 1.30. Typical UV adsorption spectra.

Typical PL spectra measured by Accent RPM2000 with an excited wavelength of 266nm are shown in Figure 1-31. It could be seen that all the TiO_2 nanoparticles exhibit a strong PL signal at the wavelength range from 370 to 450nm, with a wide peak signal from 400 to 420nm, resulting mainly from band edge free excitons, reflecting significant oxygen vacancies on the surface of TiO_2 nanoparticles [10].

Correlations of rutile fraction and the PL intensity at the wavelength of 400–420 nm of the TiO_2 nanoparticles are plotted in Figure 1-32. It could be seen that PL intensity increases as rutile mass fraction increases up to about 0.4, and decreases as rutile fraction above 0.4, reflecting electrons from valence to conductive bands more vigorously, and electronic coupling in the contacts between anatase and rutile phases [11].

1.4. SnO_2 Nanoparticles by Turbulent Diffusion Flame

Like TiO_2, SnO_2 is also an n-type semiconductor and has been used intensively for gas sensor, transparent conductor, nanoeletronic device and oxidation catalyst [12]. Photocatalytic studies on TiO_2/SnO_2 systems have been reported and the system enhances photocatalytic activity by charge separation mechanism, as the CB edges of rutile SnO_2 is about +0.0eV versus normal hydrogen electrode (NHE) at pH 7, lower than that of TiO_2, and the VB edges of rutile SnO_2 is about 3.6eV versus normal hydrogen electrode (NHE) at pH 7, higher than that of TiO_2 [13,14]. Recent results in photodegradation of methylene blue (MB) has shown that rutile SnO_2 made by thermal decomposition of tin chloride pentahydrate ($SnCl_4 \cdot 5H_2O$) has similar catalytic activity with TiO_2 phototcatalyst (P25) [15].

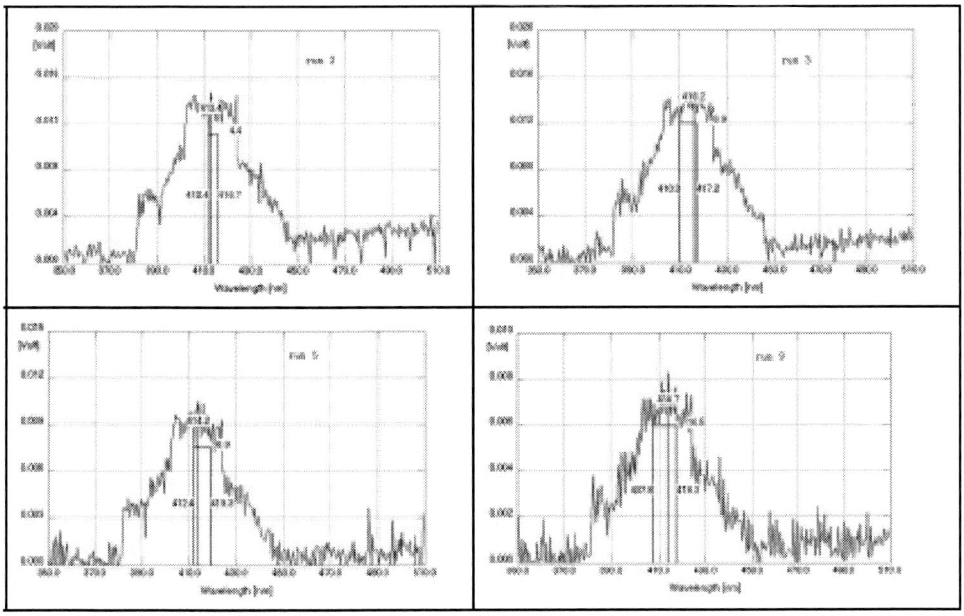

Figure 1.31. PL spectra measured with an excited wavelength of 266nm.

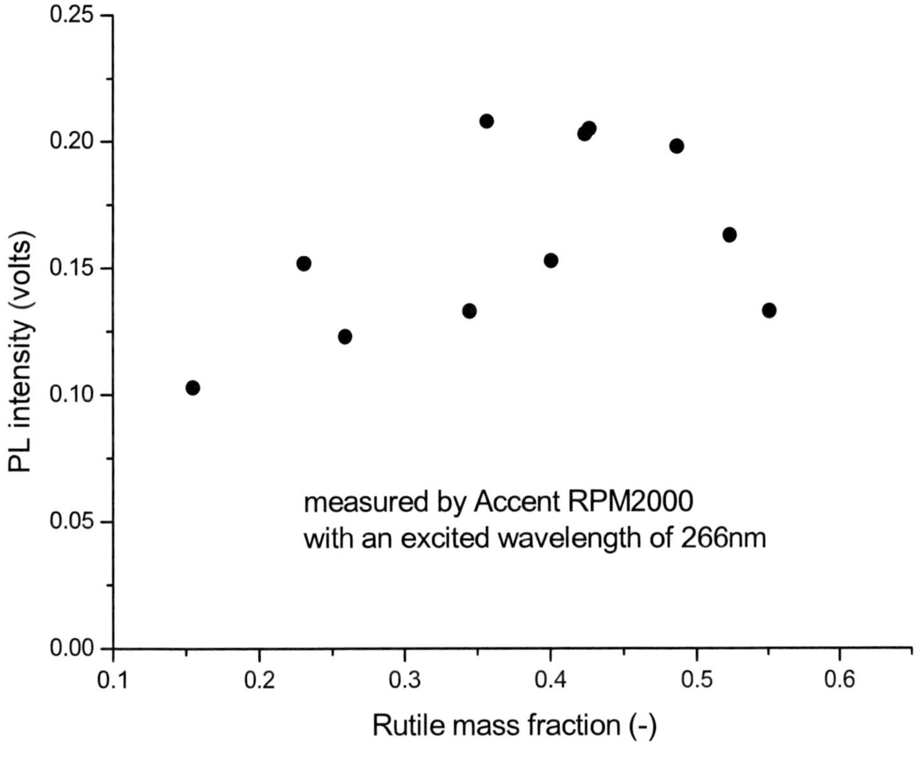

Figure 1.32. PL intensity *versus* rutile mass fraction.

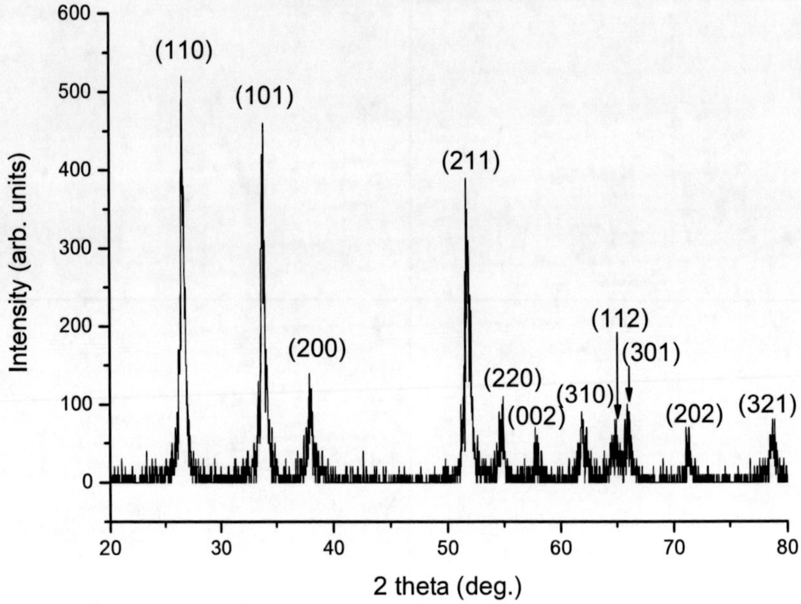

Figure 1.33. XRD pattern of SnO_2 naoparticles by turbulent diffusion flame.

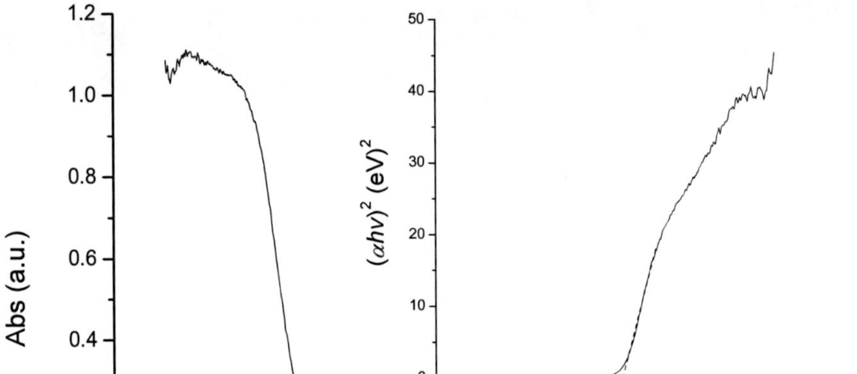

Figure 1.34. UV absorption spectrum and $(\alpha h\nu)^2$ *versus* photon energy $h\nu$ curve.

SnO_2 nanoparticles are synthesized by the oxidation of anhydrous stannic chloride ($SnCl_4$) in propane/air turbulent flame under the conditions of the flow rates of primary air of 120slm, propane of 5.577slm and carrier air for $SnCl_4$ of 30slm.

Spherical SnO_2 nanoparticles with mean size about 30nm are prepared by turbulent diffusion flame CVD process. The XRD pattern indicates that the SnO_2 nanoparticles have a tetragonal rutile structure as shown in Figure 1-33. UV absorption spectrum is shown in Figure 1-34.

As a crystalline semiconductor, the optical absorption near the band edge follows the formula $(\alpha h\nu)^n = B(h\nu - E_g)$, where α, h, ν, E_g, and B are the absorption coefficient, Planck constant, light frequency, band gap energy, and a constant, respectively [16]. Among them, n depends on the characteristics of the transition in a semiconductor, which is either 2 for direct inter-band transition or 1/2 for indirect inter-band transition. The inset of Figure 1-34 shows the $(\alpha h\nu)^2$ versus photon energy $h\nu$ curve. The band gap energy found by extrapolating the straight portion of $(\alpha h\nu)^{1/2} - h\nu$ plot to the point $\alpha = 0$ is about 3.68 eV, which is slightly higher than the reported value of 3.6eV for SnO_2 [12,14,17].

REFERENCES

[1] Pratsinis S.E. (1998). Flame aerosol synthesis of ceramic powders. *Prog. Energy Combust. Sci.*, 24, 200.

[2] Geldat D. (1986). *Gas Fluidization Technology* (pp.14-16). John Wiley and Sons Ltd.

[3] Machenzie K. (1975). The calcinations of titania: the effect of additive on the anatase-rutile transformation. *Trans. J. Br. Ceram. Soc.*, 74, 29.

[4] Spurr A., Myers H. (1957). Quantitative analysis of anatase-rutile mixtures with an X-Ray diffractometer", *Anal. Chem.*, 29, 760.

[5] Tian D., Yu G., Zhang P.Y., et. al. (2002). Photocatalytic degradation of toluene in the gas phase under carbon black modified titanium dioxide. *J. Solar Energy*. 23, 1-4 (in Chinese).

[6] Tsumura T., Kojitani N., Umemura H., Toyoda M. and Inagaki M. (2002). Composites between photoactive anatase-type TiO_2 and adsorptive carbon. *Appl. Surf. Sci.*, 196, 429-436.

[7] Sunderland P.B., Faeth G.M. (1996). Soot formation in hydrocarbon/air laminar jet diffusion flames. *Combustion and Flame*, 105, 132-146.

[8] Xie H.Y., Chen S. (2010). Simulation of combustion process and soot yield in propane/air laminar diffusion flame by self-defined scalar method. *Int. J. Chem. Reactor Eng.*, 8, Article A159.

[9] Jing L., Sun X., Xin B., Wang B., Cai W. and Fu H. (2004). The preparation and characterization of La doped TiO_2 nanoparticles and their photocatalytic activity. *J. Solid Chem.*, 177, 3375.

[10] Morooka S., Yasutake T., Kobata A., Ikemizu K. and Kato Y. (1989). A mechanism for the production of ultrafine particles of TiO_2 by a gas-phase reaction. *International Chemical Engineering*, 29, 119.

[11] Baiju, K.V., Zachariah, A., Shukla, S., Biju, S., Reddy, M.L.P., and Warrier, K.G.K. (2009). Correlating photoluminescence and photocatalytic activity of mixed-phase nanocrystalline titania. *Catalysis Letters*, *130*, 130.
[12] Batzill M., Diebold U. (2005). *Prog. Surf. Sci.* 79, 47.
[13] Vinodgopal K., Bedja I., Kamat P.V. (1996). *Chem. Mater.* 8, 2180.
[14] Levy B., Liu W., Gilbert S. (1997). Directed photocurrents in nanostructured TiO2/SnO2 heterojunction diodes.*J. Phys. Chem. B* 101, 1810.
[15] Solis-Casadors D., Vigueras-Santiago E., Hernandez-Lopez S., Camacho-Lopez M.A. (2009). *Ind. Eng. Chem. Res.,* 48, 1249.
[16] Butler M. A. (1977). *J. Appl. Phys. 48,* 1914.
[17] Nasr C., Kamat P.V., Hotchandani S. (1998). *J. Phys. Chem. B* 102, 10047.

Chapter 2

DYNAMICS OF AEROSOL PARTICLES

2.1. COALESCENCE OF AEROSOL PARTICLES

In aerosol processes, particles form by gas-phase chemical reaction, homogeneous nucleation, heterogeneous condensation, and coagulation, producing spherical primary particles. These primary particles grow into spherical or angular secondary particles by collision and coalescence, and the secondary particles may further form dendritic or fractal-like aggregates by the collisions and coalescence between secondary particles, and between primary and secondary particles, or between the aggregates. The coalescence, the sintering process in the field of powder metallurgy, is referred to the approach of colliding particles to spherical shape by viscous flow and solid-state diffusion in aerosol processes. Collisions increase the mass of a particle or aggregate, and coalescence changes their shape toward the equilibrium spherical morphology. The size and shape of the secondary particles as well as the fractal-like aggregates depend then upon both the collision rates and the rate of coalescence, and their residence time in a particular aerosol process as well.

Frenkel [1] showed that the time of coalescence depends on the viscosity, particle size, and surface tension of the material for viscous flow sintering. There are several mechanisms that govern mass transport during solid-state sintering: surface, grain boundary, volume or lattice diffusion, evaporation-condensation, and plastic flow.

Rates of neck growth, length shrinkage, and surface area reduction are vital important in determining sintering process of coalescing particles and have received extensive studies. The exponential decay of the surface area of coalescing particles obtained by Koch and Friedlander [2] has been incorporated in aerosol dynamics models to predict particle growth of nanoparticles in aerosol processes [3-5].

For the initial stage of coalescing, Frenkel [1] obtained a power law $X \propto t^{1/2}$ for the growth of the neck size of coalescing particles, by numerical analysis of the coalescence process from viscous flow theory. Hiram and Nir [6] found that the neck size X approaches its final value X_f exponentially as

$$\frac{dX}{dt} \propto -\frac{1}{\tau}\left(X - X_f\right) \qquad (2\text{-}1)$$

where τ is the characteristic coalescence time, and X is the neck size of coalescing particles. From solid-state diffusion driven by stress gradients resulting from nonsphericity, Friedlander and Wu [7] obtained a linear rate law for the decay of the surface area of coalescing particles as

$$\frac{da}{dt} \propto -\frac{1}{\tau}\left(a - a_{sph}\right) \tag{2-2}$$

where a_{sph} is the surface area of the final sphere. This linear rate law was also proposed by Koch and Friedlander [2] from the Hiram and Nir relation (2-1), and similar results were also obtained by Yadha and Helble [8] from viscous theory. The linear rate law is now frequently used in the modeling of the growth of nanoparticles in aerosol processes.

Studies have shown that rates of neck growth, length shrinkage, and surface area reduction of coalescing particles could be approached with geometrical evolution by geometrical models [8-10] incorporating with rate expressions of viscous flow or solid-state diffusion mechanisms. The geometrical evolution of two coalescing particles has been modeled by using the form of Hippopede curves together with the condition of continuity or mass conservation throughout coalescence process. In these models, the coalescence process is characterized by a coalescence parameter, which is an adjustable parameter in the form of a Hippopede curve. [8,9]

In this book, the geometrical evolution of two coalescing particles is modeled by a geometrical model under the assumption that the mass and density of each particle remain unchanged and each particle is in the shape of a spherical cap during coalescing. The rates of neck size growth, decay of surface area and shape evolution of the coalesced particle are calculated from the Fick law for solid-state diffusions of volume, grain boundary and surface.

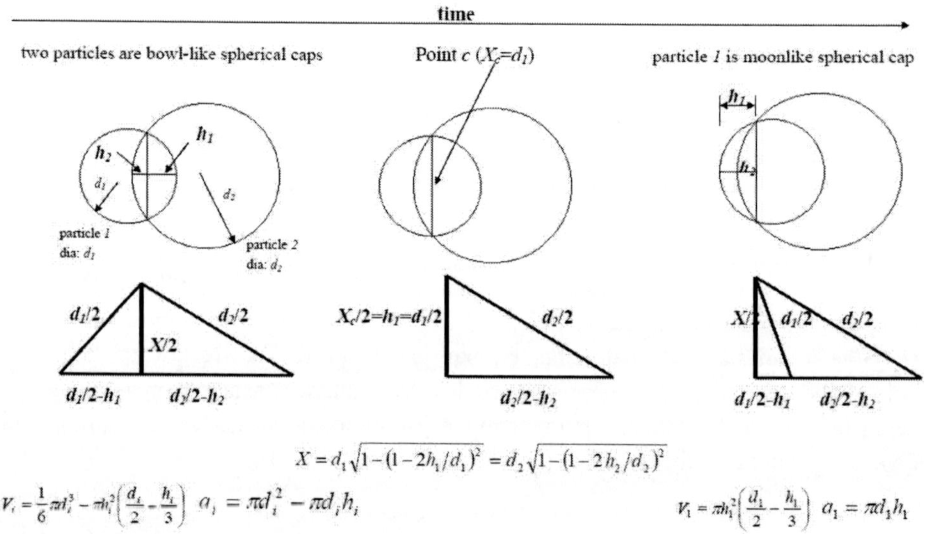

Figure 2.1. Illustration of geometrical relations of two coalescing particles.

2.1.1. Geometrical Model for Two Coalescing Particles

Under the assumptions that the mass and density of each particle remain unchanged and each particle is in the shape of a spherical cap during coalescing, the coalescence process could then be characterized directly by the neck size of two coalescing particles as shown in Figure 2-1.

At the beginning of coalescence, the two spheres just come into contact, the neck size is zero and the coalesced cap height of particle *1* or particle *2* is also zero. As time proceeds, the neck size grows, the two particles become bowl-like spherical caps and their coalesced caps are moonlike spherical caps, the diameters of their father spheres and their coalesced cap heights increase.

When the neck size becomes to equal the diameter of its father sphere of the small particle, the small particle becomes a semi-sphere and the coalesced cap is the other semi-sphere, which is set as point *c* in Figure 2-1.

As time further proceeds, the large particle is still a bowl-like spherical cap and its coalesced cap is still a moonlike spherical cap, but the small particle becomes a moon-like spherical cap and its coalesced cap becomes a bowl-like spherical cap and the diameter of its father sphere becomes larger than the neck size until the diameters of the two father spheres are equal, the two particles are merged and the coalescence is finished.

From the geometrical relationships in Figure 2-1, where particle *1* is smaller than particle *2*, the neck diameter X in the entire coalescing process is equal to

$$X = d_1\sqrt{1-(1-P_1)^2} = d_2\sqrt{1-(1-P_2)^2} \tag{2-3}$$

where

$$P_i = \frac{2h_i}{d_i} = 1-\sqrt{1-(X/d_i)^2} \tag{2-4}$$

where d_1 and d_2 are the diameters of sphere *1* and *2*, respectively, h_1 and h_2 are the heights of the coalesced caps of particle *1* and *2*, respectively.

Before point *c*, since both two particles are bowl-like spherical caps, the volume of particle *1* or *2* is then equal to

$$V_i = \frac{1}{6}\pi d_i^3 - \pi h_i^2\left(\frac{d_i}{2}-\frac{h_i}{3}\right) \tag{2-5}$$

Since solid diffusion is happened in the neck region, it is assumed that the volume of a particle is not changed during the coalescing, e.g. $V_1 = V_{10}$ and $V_2 = V_{20}$, then it can be obtained from equations (2-3) to (2-5) that

$$d_i^3 = \frac{4d_{i0}^3}{4 - 3P_i^2 + P_i^3} \tag{2-6}$$

where d_{i0} is the initial diameter of sphere *1* or *2*.

The surface area of particle *1* or *2* is equal to

$$a_i = \pi d_i^2 - \pi d_i h_i = \pi d_i^2 (1 - P_i/2) \tag{2-7}$$

the surface area of the coalesced particle a is then equal to

$$a = a_1 + a_2 = \pi d_1^2 (1 - P_1/2) + \pi d_2^2 (1 - P_2/2) \tag{2-8}$$

The sphericity of the coalesced particle ψ, which is defined by the ratio of the surface area of a sphere having the same volume to the surface area of the particle, is equal to

$$\psi = \frac{\left(d_{10}^3 + d_{20}^3\right)^{2/3}}{d_1^2(1 - P_1/2) + d_2^2(1 - P_2/2)} \tag{2-9}$$

The surface area and sphericity of the coalesced particle can therefore be calculated from the neck size and initial diameters of the two spheres before point c.

At point c, from equations (2-4) and (2-6)

$$X_c = d_{1c} = \sqrt[3]{2} d_{10} = \alpha \sqrt[3]{2} d_{20} \tag{2-10}$$

where α is the ratio of the initial diameters d_{10}/d_{20}.

After point c, since particle *2* is still a bowl-like spherical cap, equations (2-5) and (2-7) are still hold for its volume and surface area, respectively. But particle *1* becomes a moon-like spherical cap and its volume is then equal to

$$V_1 = \pi h_1^2 \left(\frac{d_1}{2} - \frac{h_1}{3}\right) = \frac{\pi}{6} d_1^3 P_1^2 \left(1 - \frac{P_1}{3}\right) \tag{2-11}$$

The diameter of its father sphere can then be obtained as

$$d_1^3 = \frac{4d_{10}^3}{P_1^2 (3 - P_1)} \tag{2-12}$$

The surface area of particle *1* is equal to

$$a_1 = \pi d_1 h_1 = \pi d_1^2 P_1/2 \tag{2-13}$$

The surface area of the coalesced particle is then equal to

$$a = \pi d_1^2 P_1/2 + \pi d_2^2 (1 - P_2/2) \qquad (2\text{-}14)$$

The sphericity of the coalesced particle equals then

$$\psi = \frac{\left(d_{10}^3 + d_{20}^3\right)^{2/3}}{d_1^2 P_1/2 + d_2^2 (1 - P_2/2)} \qquad (2\text{-}15)$$

At the final coalescence, the two particles form one particle and the diameter of the coalesced particle is equal to

$$d_f = d_{1f} = d_{2f} = \sqrt[3]{d_{10}^3 + d_{20}^3} = \sqrt[3]{1 + \alpha^3}\, d_{20} \qquad (2\text{-}16)$$

where the subscript f stands for the final coalescence. From equations (2-12) and (2-16), it can be obtained that

$$P_f^2 (3 - P_f) = \frac{4\alpha^3}{1 + \alpha^3} \qquad (2\text{-}17)$$

From equation (2-4), the neck size at the final coalescence is equal to

$$X_f = d_f \sqrt{1 - (1 - P_f)^2} \qquad (2\text{-}18)$$

Thus, the surface area and sphericity of the coalesced particle can be calculated by given the neck size and initial diameters of the two particles in the entire coalescing process.

The initial length of coalescing particles, L_0, is

$$\mathrm{L}_0 = \mathrm{d}_{10} + \mathrm{d}_{20} \qquad (2\text{-}19)$$

and the length of coalescing particles, L, is

$$\mathrm{L} = (\mathrm{d}_1 - \mathrm{h}_1) + (\mathrm{d}_2 - \mathrm{h}_2) \qquad (2\text{-}20)$$

The Shrinkage length of coalescing particles, ΔL, is defined as

$$\Delta \mathrm{L} = \mathrm{L}_0 - \mathrm{L} \qquad (2\text{-}21)$$

The calculated result of normalized shrinkage length for coalescing spheres is compared with that obtained by the Yadha and Helble model [4] in Figure 2-2 against normalized neck size. The shrinkage length and the neck size are normalized by the diameter of the smaller particle. Present model and the Garabedian and Helble model [9] is compared in Figure 2-3 for equal sized coalescing spheres, where the shrinkage length is normalized by the diameter of the smaller particle, and the neck size is normalized by the neck size at the final coalescing, X^*. The figures show that the models give similar results in the entire coalescing process. The present model has the advantage of simplicity.

The calculated results of normalized excess surface area and sphericity of coalescing particles are plotted against the normalized neck size in Figures 2-4 and 2-5, respectively. The excess surface area is normalized by the surface area of the final coalesced sphere. Figures 2-4 and 2-5 show that the excess surface area decreases to zero and the sphericity increases to unity when the neck size approaches to its final value. The sphericity and the normalized excess surface area versus the normalized neck size depend only upon the ratio of the initial size of the two particles α, but are independent on their initial values.

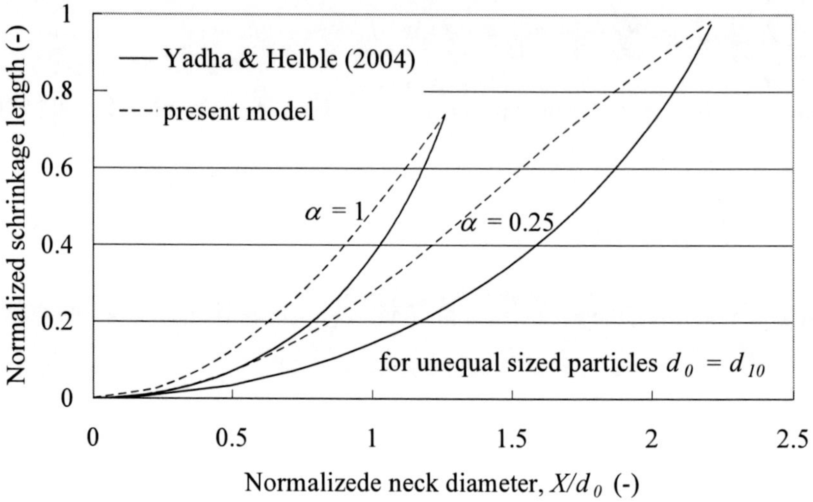

Figure 2.2. Normalized shrinkage length of coalesced particle *versus* normalized neck size, X/d_0.

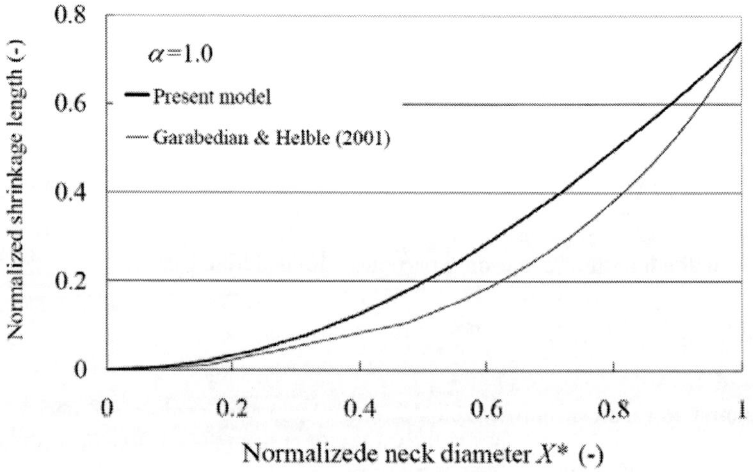

Figure 2.3. Normalized shrinkage length of coalesced particle *versus* normalized neck size, X/X_f.

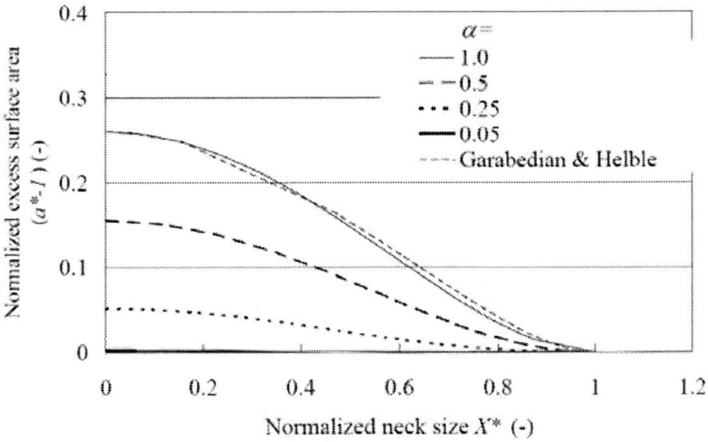

Figure 2.4. Normalized excess surface area of coalesced particle *versus* normalized neck size.

The calculated results of the normalized excess surface area and sphericity of two equal sized coalescing spheres by the Garabedian and Helble's model [9] are also presented in Figures 2-4 and 2-5, respectively. The results show that the two models give similar results in the entire coalescing process for equal sized coalescing particles.

2.1.2. Coalescencing Regimes

For the coalesce of two spheres, the coalesced happens around the neck region as observed by Nichols and Mullins, [11] the radius of the neck surface ρ illustrated by Figure 2-6 is equal to

$$\rho = \frac{d_2}{2}(1-\cos\theta) = \frac{d_2}{2}\left(1-\sqrt{1-X^2/d_2^2}\right) = \frac{1}{2}d_2 P_2 \tag{2-22}$$

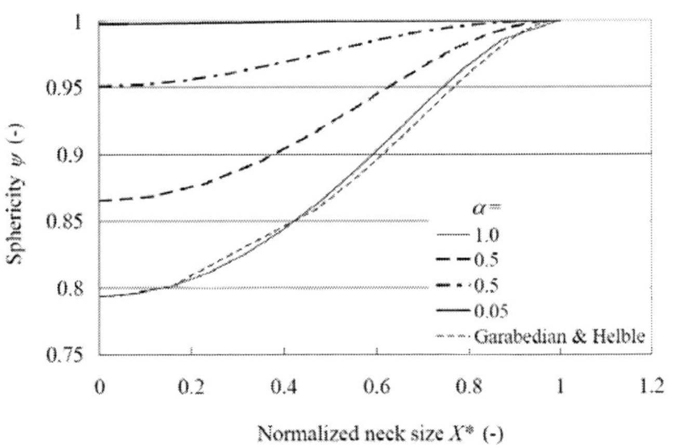

Figure 2.5. Sphericity of coalesced particle *versus* normalized neck size.

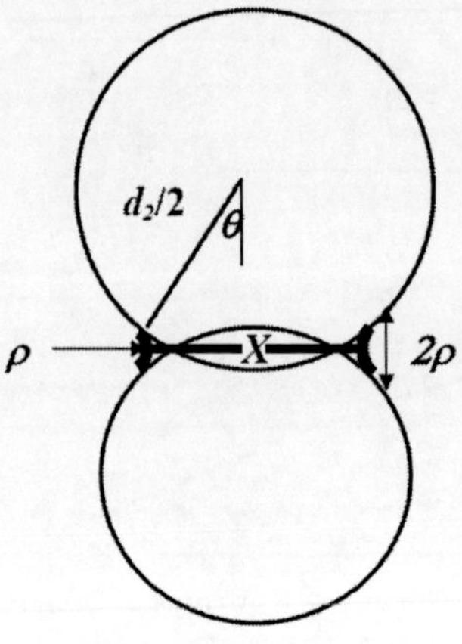

Figure 2.6. Illustration of neck surface in two coalescing particles.

the volume and surface area of the neck are then equal to

$$V_n = \frac{\pi}{4} X^2 \rho = \frac{\pi}{8} X^2 d_2 P_2;\ A = \pi X \rho = \frac{\pi}{2} X d_2 P_2 \qquad (2\text{-}23)$$

Then by using Fick's laws for volume, grain boundary and surface diffusion, [12,13] one can obtained that

$$\frac{dF}{dt} = \begin{cases} K_v \dfrac{X}{d_2 P_2} & \text{for volume diffusion} \\ K_{gb} \dfrac{1}{d_2 P_2} & \text{for grain boundary diffusion} \\ K_s \dfrac{X}{d_2 P_2} & \text{for surface diffusion} \end{cases} \qquad (2\text{-}24)$$

where

$$F = X^2 d_2 P_2 \qquad (2\text{-}25)$$

K_v, K_{gb} and K_s are the rate constants of volume, grain boundary and surface diffusion, respectively.

The neck size X could then be calculated from equations (2-24) and (2-25) together with the relationships obtained by kinematics at given initial particle sizes and the rate constants K_v, K_{gb} and K_s. Such calculated results of the normalized change rate dX^*/dt^* versus the normalized neck size X^* is shown in Figure 2-7 at different α values from 0.05-1.0 in the entire coalescing process.

The figure shows that the normalized change rate dX^*/dt^* is inverse proportional to the normalized neck size X^* in the power from 4.0-4.4 for volume diffusion (Figure 2-7a), which is very close to that obtained by Kuczynski [13] for the initial stage of coalescing; in the power from 5.0-5.4 for grain boundary diffusion (Figure 2-7b), which is very close to that for the initial stage of coalescing given by Guo [12]; and in the power from 6.0-6.5 for surface diffusion (Figure 2-7c).

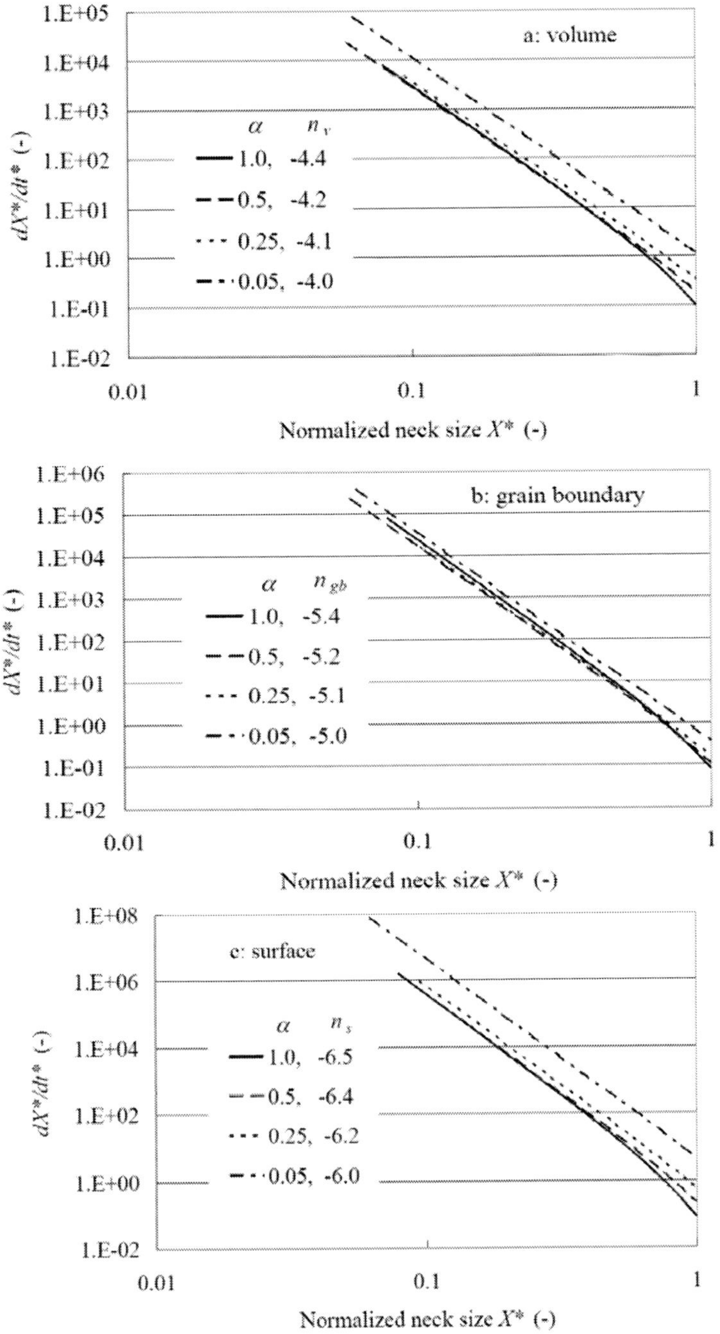

Figure 2.7. Normalized change rate dX^*/dt^* *versus* normalized neck size X^*.

In Figure 2-7, the neck size X is normalized by the neck size at the final coalescence X_f, and the time t is normalized by the characteristic coalescence times of volume diffusion τ_v, grain boundary diffusion τ_{gb} and surface diffusion τ_s, respectively, and they are

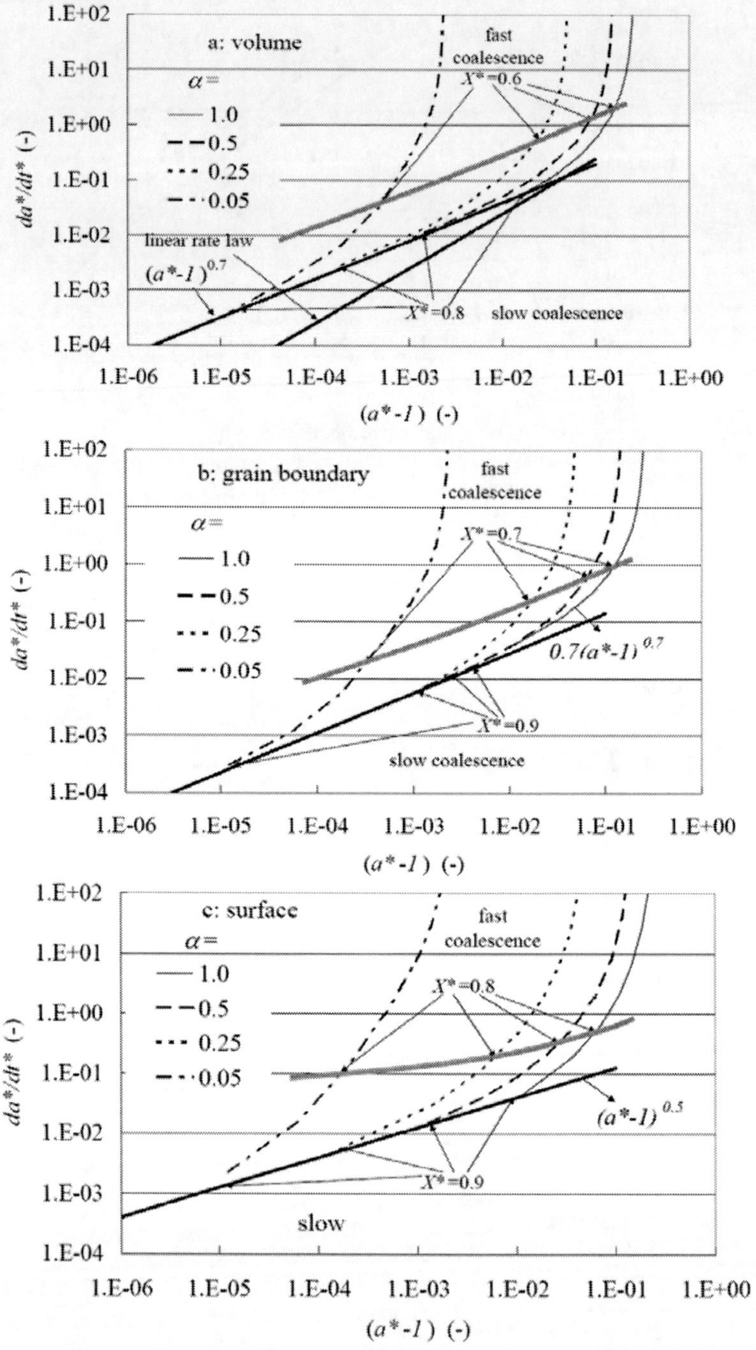

Figure 2.8. Normalized decay rate of surface area *versus* normalized excess surface area.

$$\tau = \begin{cases} \dfrac{d_{10}^{3}}{K_v} & \text{for volume diffusion} \\ \dfrac{d_{10}^{4}}{K_{gb}} & \text{for grain boundary diffusion} \\ \dfrac{d_{10}^{4}}{K_s} & \text{for surface diffusion} \end{cases} \tag{2-26}$$

The normalized decay rates of surface area versus normalized excess surface area are plotted in Figure 2-8 at the different α values.

The figure indicates that coalescence is very fast before the neck size grows up to about 60% of its final value for volume diffusion (Figure 2-8a), 70% for grain boundary diffusion (Figure 2-8b) and 80% for surface diffusion (Figure 2-8c) in a time less than about 10% of their characteristic coalescence times; this regime is called fast coalescence regime. And the coalescence becomes slowly and finally the surface area decays at an asymptotic rate, after the neck size grows above 80% of its final value for volume diffusion (Figure 2-8a), 90% for grain boundary (Figure 2-8b) and surface diffusion (Figure 2-8c), as

$$\frac{da^*}{dt^*} = \begin{cases} (a^*-1)^{0.7} & \text{for volume and grain boundary diffusion} \\ (a^*-1)^{0.5} & \text{for surface diffusion} \end{cases} \tag{2-27}$$

From the rate constant K_v of volume diffusion [12] and the characteristic frequency factor,[7] the linear rate law equation (2-2) should be

$$\frac{da^*}{dt} = 2.5(a^*-1) \tag{2-28}$$

It can be seen from Figure 2-8a that the asymptotic coalescence rate is slightly different from the linear law obtained from stress gradients by Friedlander and Wu [7] as given in equation (2-2). This may be caused by the assumption that the mass and density of each particle remain unchanged and each particle is in the shape of a spherical cap during coalescing.

The calculated evolution of the sphericity of the coalescing particles is presented in Figure 2-9. It can be seen that thc coalescing particles has been growing into one particle even before the end of the fast coalescence regime, and has developed into a particle in angular shape before the slow coalescence regime.

The results may yield that surface diffusion is the fastest process in particle coalescing among the solid diffusions, and the fast coalescence regime is ended at the neck size up to about 80% of its final value, e.g., the coalesced particle can grow into one particle in a short time.

From Figure 2-8c it can be seen that the time for the neck size growing up to 80% of its final value is ranging from 0.01 to 0.16τ_s corresponding to the α value from 0.05-1.0.

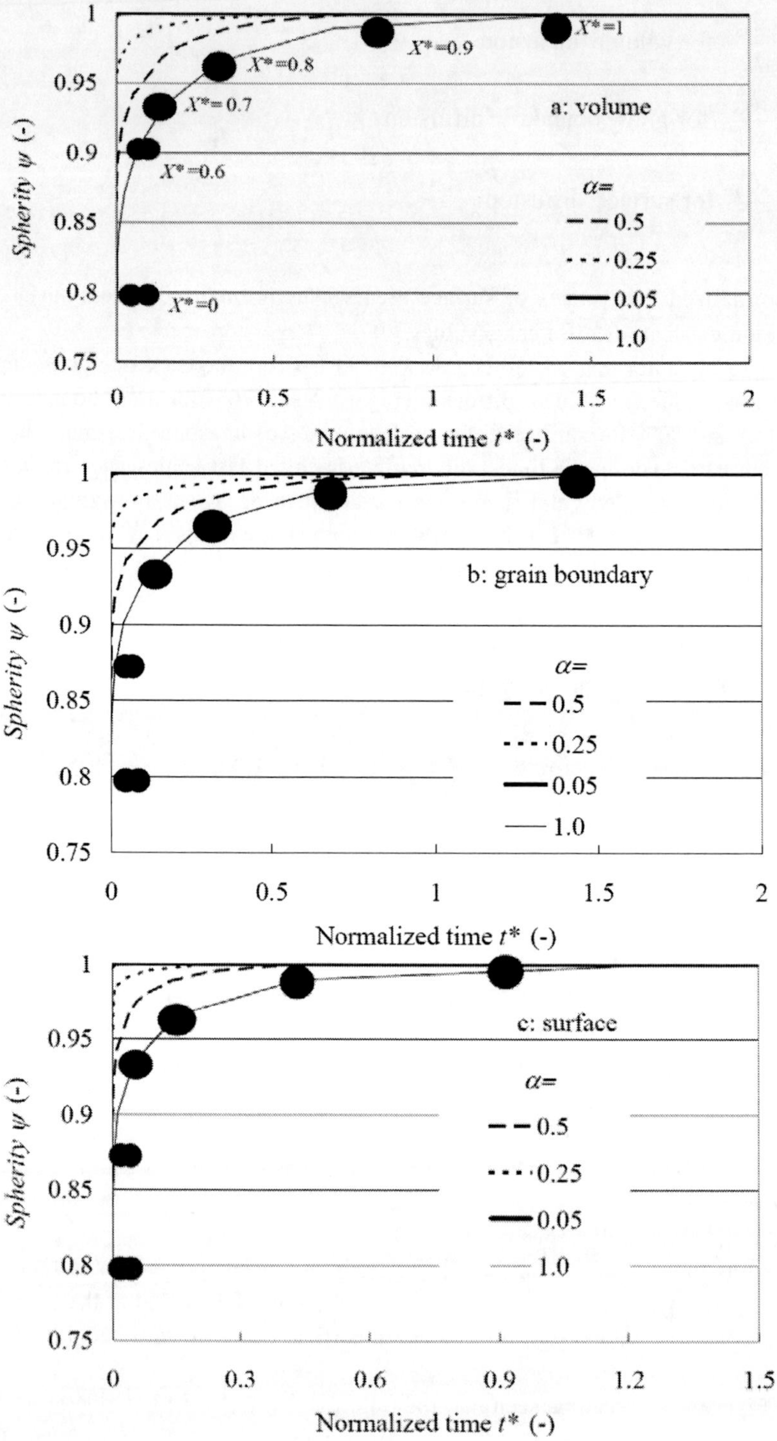

Figure 2.9. Sphericity of coalescing particles *versus* normalized time t^*.

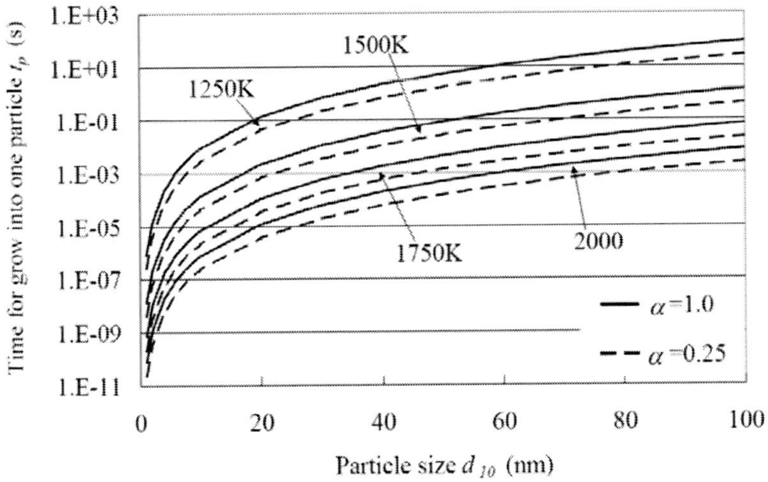

Figure 2.10. Time for coalescing particles growing into one particle t_p *versus* particle size at different temperatures.

2.1.3. Coalescing by Surface Diffusion Mechanism

For titania particles, the surface diffusion rate constant K_s could be given by Kobata et al., [14]

$$K_s = k_0 e^{-\frac{E_s}{RT}} \tag{2-29}$$

where k_0 is the frequency factor and is $1.143 \cdot 10^{-21} m^4/s$, E_s is the activation energy of surface diffusion and is 258kJ/mol, and R is the gas constant. The time for the coalesced particle growing into one particle t_p can then be obtained from equations (2-26) and (2-29) and such calculated results are shown in Figure 2-10.

It can be seen that the time for the coalesced particle growing into one particle t_p strongly depends on particle size and temperature, and could be a few *ns* or *μs* for particles less than 20nm and could be a few hundred *ms* for larger particles or at low temperatures. Since the dendritic aggregates could be formed only when particle residence time is much smaller than the time for the coalesced particle growing into one particle t_p. It is, therefore, the dendritic aggregates could be only formed from secondary particles in a few tens of nanometers for titania synthesized in flame CVD process.

2.1.4. Coalescing by Volume Diffusion Mechanism

The calculated results of normalized excess surface area, $(a - a_f)/(a_0 - a_f)$, for two equal-sized coalescing spheres by present model, the Yadha and Helble model, and the Koch and Friedlander model are compared against normalized coalescing time in Figure 2-11 based on volume solid-state diffusion mechanism. The results show that the three models give similar prediction. Present model predicts slightly fast coalescing process than the other two models. This may be again due to the assumption of spherical-cap shape of coalescing particles. The rate expression for volume solid-state diffusion used in the calculations is [12]

$$\left(\frac{X}{d}\right)^5 = \frac{320\gamma D_v\Omega}{kTd^3}t,\ \tau_v = \frac{kTd_0^3}{320\gamma D_v\Omega} \tag{2-30}$$

where k is the Boltzmann constant (J/K), T is the absolute temperature (K), D_v is volume diffusion coefficient (m^2/s), γ is the surface tension (J/m^2), Ω is the atomic volume (m^3), τ_v is the characteristic coalescing time (s).

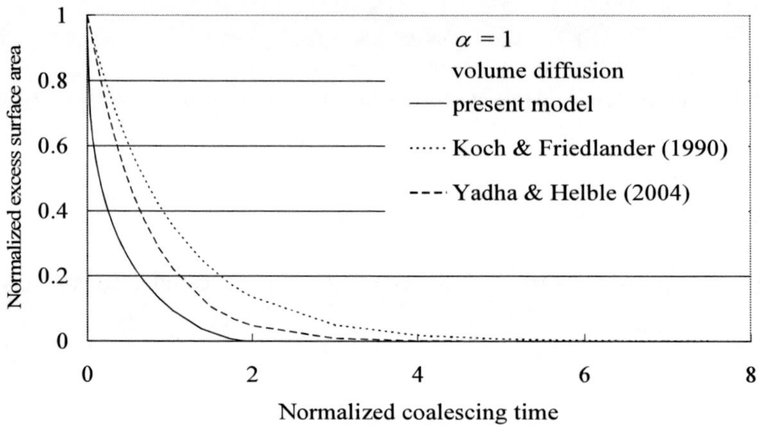

Figure 2.11. Predicted normalized excess surface area *versus* normalized coalescing time for equal-sized spheres.

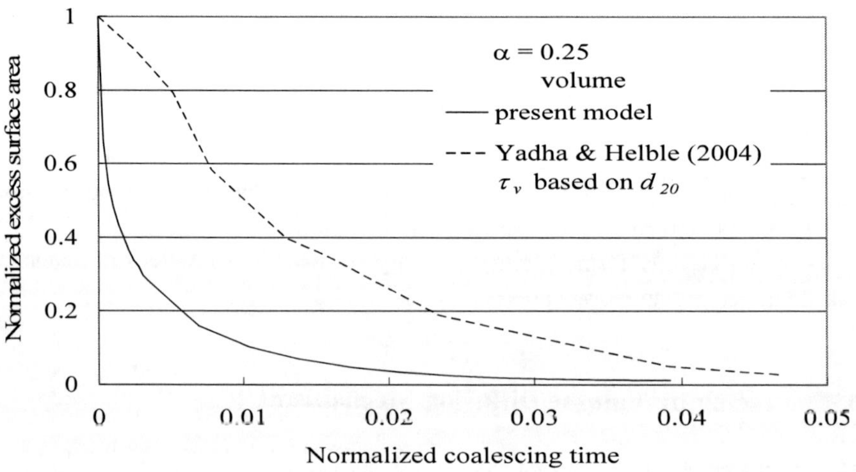

Figure 2.12. Predicted normalized excess surface area *versus* normalized coalescing time for unequal-sized spheres.

The calculated results of normalized excess surface area, $(a - a_f)/(a_0 - a_f)$, for two unequal-sized coalescing spheres by present model and the Yadha and Helble model are compared in Figure 2-12 against normalized coalescing time based on volume solid-state diffusion mechanism for the case that the diameter of the large particle is four times of that of the small particle ($\alpha = 0.25$). In the calculations, the characteristic coalescing time is based on the initial diameter of the large particle. The results show that present model predicts fast coalescing process than the Yadha and Helble model, due to the assumption of spherical-cap shape of coalescing particles, but the two models give similar results.

Figure 2-13 shows the calculated results of normalized neck size, diameters of the small and large particles in coalescing process for the case in Figure 2-12. It can be seen that the large particle increases lightly as time proceeds, and normalized diameter of the large particle increases up to 1.005 (d_{2f}/d_{20}) at the final coalescing which can be also obtained by equation (2-16). The normalized diameter of the small particle increases almost linearly with normalized coalescing time and up to 1.005 at the final coalescing (d_{1f}=d_{2f}). The normalized neck size increases fast before point *c*, at point *c* X_c is equal to d_{1c}, and increases slowly after point *c* and up to 0.524 (X_f/d_{20}=0.524, or X_f/d_{10}=1.26) at the final coalescing.

2.1.5. Coalescing by Overall Diffusion Mechanism

Predictions on normalized excess surface area by present model and the Koch and Friedlander model are compared in Figure 2-14 with literature data of titania sintering at 1173K [15] based on volume and overall solid-state diffusion mechanisms. The rate expression for volume solid-state diffusion and the characteristic coalescing time τ_v is given by equation (2-27) [12]. The rate expression for overall solid-state diffusion and the characteristic coalescing time $\tau_{overall}$ is given by [10]

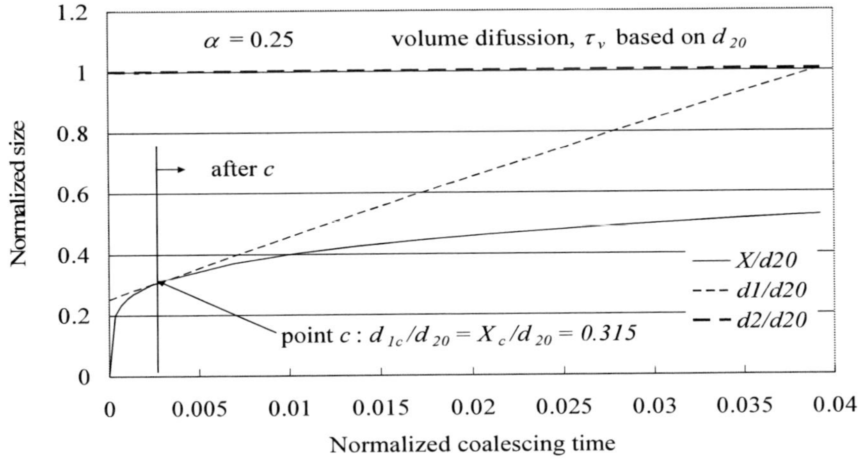

Figure 2.13. Predicted neck, and particle sizes in the entire coalescing process.

$$X^2 = \frac{64\gamma D\Omega}{kTd}, \quad \tau_{\text{overall}} = \frac{kTd_0^3}{64\gamma D\Omega} \tag{2-31}$$

where D is overall diffusion coefficient (m^2/s), and $\tau_{overall}$ is overall diffusion characteristic coalescing time (s).

In the calculations, the surface tension γ is taken as $0.6 J/m^2$ and the atomic volume Ω is taken as 1.57×10^{-29} m^3 [10]. The volume diffusion coefficient D_v (m^2/s) is taken as $1.6\times10^{-4}\exp(-310.1 kJmol^{-1}/RT)$ [15], and the overall diffusion coefficient D (m^2/s) is taken as $1.086\times10^{-2}\exp(-323 kJmol^{-1}/RT)$ [10,16]. The primary particle size of the titania is 11nm and absolute temperature T is 1173K [15].

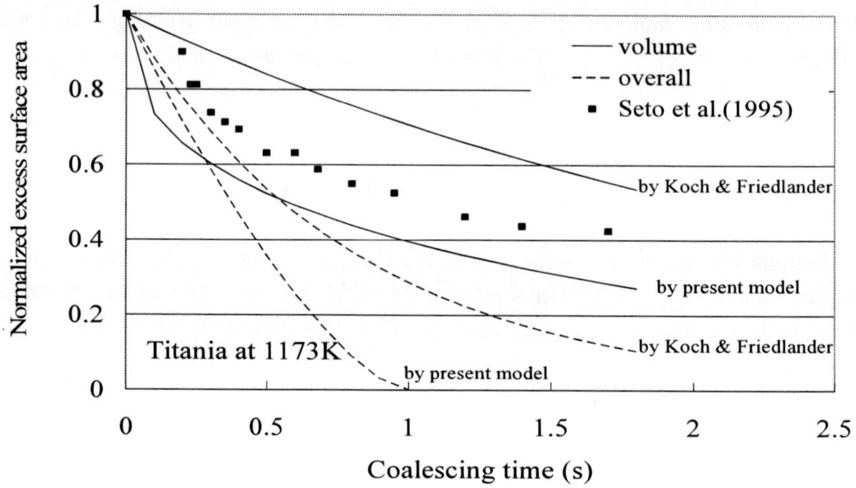

Figure 2.14. Comparison on excess surface area between predicted and literature data.

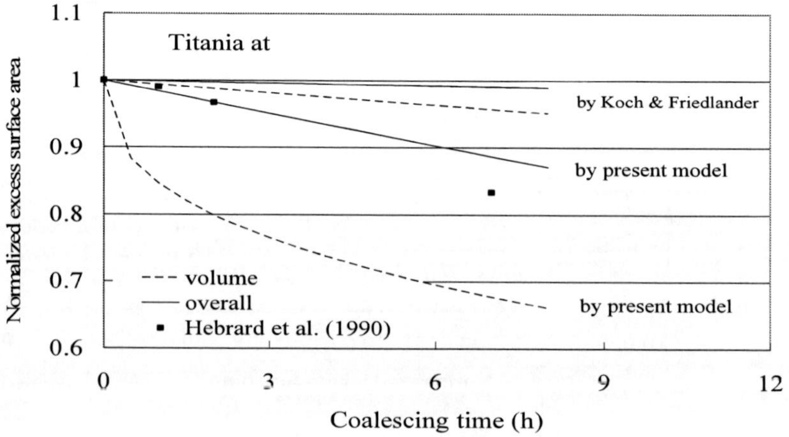

Figure 2.15. Comparison on excess surface area between predicted and literature data.

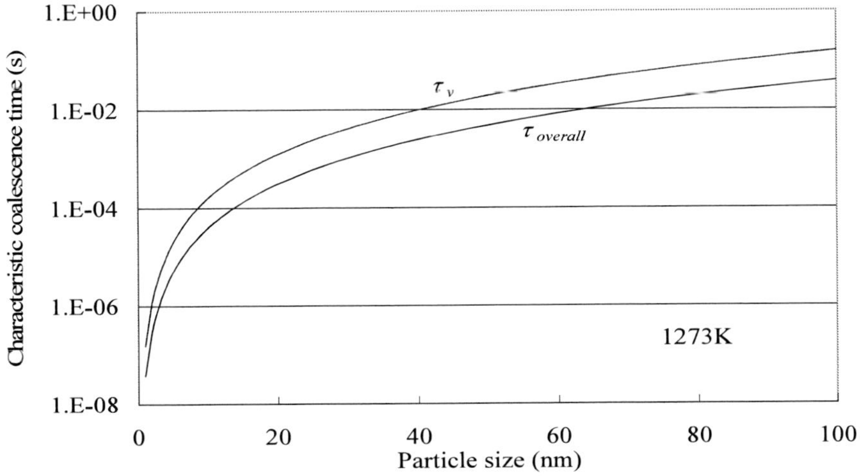

Figure 2.16. Characteristic coalescence time.

Figure 2-14 shows that both present model and the Koch and Friedlander model give reasonable predictions on excess surface area based on overall solid-diffusion mechanism at initial stage of coalescence, and based on volume solid-state diffusion mechanism at later stages of coalescing process, as reported by German [17] and Rahaman [18].

Figure 2-15 shows a comparison on excess surface area between predicted and literature data on titania sintering at a temperature of 848K [19]. The predictions are made by present model and the Koch and Friedlander model based on volume and overall solid-state diffusion mechanisms. The rate expressions and related constants used in the calculations are the same as those used in the calculations in Figure 2-14. The primary particle size of the titania is 12.3nm and absolute temperature T is 848K [19]. It can be seen that present model based on overall diffusion gives reasonable predictions at the earlier stage of coalescence.

2.1.6. Characteristic Coalescence Time

Characteristic coalescence time calculated by equations (2-30) and (2-31) is shown in Figure 2-16 for volume diffusion and overall diffusion at a temperature of 1273K, respectively. It can be seen that the characteristic coalescence time increases rapidly as particle size is less than 20nm from about 0.1μs to about 1ms, and increases more slowly when particles are more than 20nm from about 1ms to 0.1s.

Nomenclature

a surface area, m^2
a_{sph} surface area of the final coalesced spherical particle, m^2
a^* surface area normalized by a_{sph}
A surface area of neck, m^2
d particle size, m
D diffusion coefficient, m^2/s
E activation energy, J/mol

F volume parameter defined by *eq.2*-25, m^3
h height of spherical cap, m
k Boltzmann constant, J/K
k_0 frequency factor surface diffusion, m^4/s
K_v rate constant of volume diffusion, m^4/s
K_{gb} rate constant of grain boundary diffusion, m^4/s
K_s rate constant of surface diffusion, m^4/s
L length of coalescing particles, m
n index in neck change rate
P spherical cap parameter defined by *eq.2*- 6
R the gas constant, J/mol/K
t time, s
t^* time normalized by characteristic coalescence time
t_p time for coalesced particle growing into one particle, s
T temperature, K
V particle volume, m^3
V_n volume of neck, m^3
X neck size, m
X^* neck size normalized by the neck size at the final coalescing

Greet Letters

ratio of the initial size of two coalescing particles
angle defined in Figure 2-5, rad
radius of neck surface, m
characteristic coalescence time, s
Ω atomic volume, m^3
sphericity of coalesced particle

Subscripts

0 initial coalescing
for particle *1*
for particle *2*
c point for the center of particle *1* just at the center of neck
f final coalescing
gb grain boundary diffusion
i for *1* or *2*
overall overall solid difussion
s surface diffusion
v volume diffusion

2.2. Integral Collision Kernel

Various aerosol processes have been developed for the generation of nano or submicron particles at laboratory or industry scales, such as flame CVD, tube furnace, gas-condensation, thermal plasma, laser, sputtering and other aerosol processes. The growth or coagulation of particles in these processes strongly influences the particle size, size distribution and morphology, and is therefore of fundamental interest in a wide range of applications of science and engineering.

It is well accepted that when Knudsen number for particles is the order of ten or greater (e.g. for nanoparticles), the coagulation is in the free molecule regime, the collision rate and collision kernel can be evaluated from the kinetic theory of dilute gases [20]; when Knudsen number is smaller than unity, the coagulation is in the continuum regime, the collision rate and the collision kernel are obtained from Smoluchowski theory with the Stokes-Cunningham slip correction [20,21]. There is no rigorous theory for the transition regime when the particle size is comparable to the mean free path of the gas molecules. In this regime, the collision kernel is approached by semi-empirical interpolations between the two limiting regimes by a number of authors, which have been reviewed by Otto and Fissan [21].

2.2.1. Drag Force of Aerosol Particles

For small and spherical particles moving in Newton fluids, particle Reynolds number is much smaller than unity, drag force F_D obeys Stokes law as

$$F_D = 3\pi\mu du \tag{2-32}$$

where μ is fluid viscosity, u is particle velocity and d is particle diameter.

When particle size is close to the mean free path of fluid molecules, the collisions between the particle and the molecules becomes discontinuum, Stokes law should be modified by Cunningham slip factor as

$$F_D = \frac{3\pi\mu\, du}{C_C} \tag{2-33}$$

The Cunningham slip factor is

$$C_C = 1 + 2Kn\left(1.257 + 0.4e^{-0.55/Kn}\right) \tag{2-34}$$

where Kn is Knudsen number

$$Kn = \frac{\lambda}{d} \tag{2-35}$$

where λ is the mean free path of the gas molecule.

Cunningham slip factor versus Knudsen number is graphically illustrated in Figure 2-17. It shows that when Knudsen number is less than 0.1, Cunningham slip factor is about unity, the drag force F_D obeys Stokes law. When Knudsen number is more than 0.1, Cunningham slip factor increases rapidly as Knudsen number increases, Stokes law has to be modified by Cunningham slip factor.

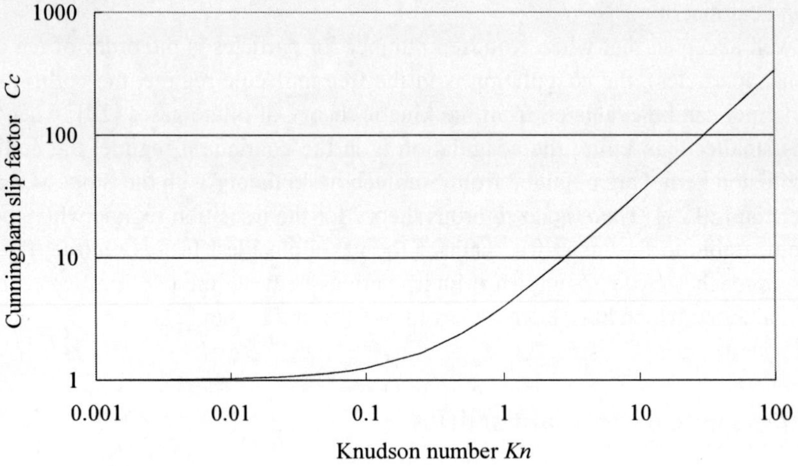

Figure 2.17. Cunningham slip factor *versus* Knudsen number.

The molecule number density of an ideal gas under standard conditions, N, is equal to

$$N = \frac{6.023 \times 10^{23}}{22.4 \times 10^{-3}} = 2.69 \times 10^{25} \tag{2-36}$$

The size of nitrogen molecules is 0.48nm, the mean free path of nitrogen molecules is

$$\lambda = \frac{1}{\sqrt{2}\pi N \upsilon_0^2} = 36.3\text{nm} \tag{2-37}$$

For nitrogen molecules under ambient pressure and 1273K temperature, the molecule number density is

$$N = \frac{6.023 \times 10^{23}}{22.4 \times \frac{1273}{273} \times 10^{-3}} = 5.76 \times 10^{24} \tag{2-38}$$

The mean free path of nitrogen molecules under ambient pressure and temperature of 1273K is then equal to

$$\lambda = \frac{1}{\sqrt{2}\pi N \upsilon_0^2} = 170\text{nm} \tag{2-39}$$

For a flame CVD process with air as oxidants, typical operational conditions are under ambient pressure and temperature of 1273K. Thus, when $Kn \leq 0.1$, *i.e.* particles are larger than 1.7μm, particles obey the Stokes law. When particles are smaller than 1.7μm, the Stokes law should be modified by Cunningham slip factor.

2.2.2. Brownian Motion of Aerosol Particles

For small particles suspended in a fluid, collisions between particles and molecules will cause the particles random motion, called Brownian motion. For the particles under Brownian motion, particles diffusion obeys Fick's law as

$$j = -D_P \frac{\partial N}{\partial x} \tag{2-40}$$

where j is particle number flux, D_P is the Stokes-Einstein diffusion coefficient

$$D_P = \frac{C_C}{3\pi\mu d} kT \tag{2-41}$$

where k is Boltzmann constant, T is thermodynamic temperature.

The relationship of the Stokes-Einstein diffusion coefficient and Knudson number is shown in Figure 2-18. It shows that the Stokes-Einstein diffusion coefficient increases rapidly as Knudson number increases. When Knudson number is more than 100, the Stokes-Einstein diffusion coefficient increases slowly and approaches gradually to the diffusion coefficient of O_2 molecules in N_2 gas at the temperature of 1273K.

Thus, regimes of motion of aerosol particles might be classified as

$$\text{Motion of aerosol particles} \begin{cases} \text{Kn} < 0.1 & \text{continuum mechanics} \\ 0.1 < \text{Kn} < 100 & \text{transi tion regime} \\ \text{Kn} > 100 & \text{molecular dynamics} \end{cases}$$

For particles in N_2 gas at ambient pressure and temperature of 1273K, regimes of motion of aerosol particles might be classified as

$$\text{Motion of aerosol particles} \begin{cases} \text{d} > 1700\,\text{nm} & \text{continuum mechanics} \\ 1.7\text{nm} < \text{d} < 1700\,\text{nm} & \text{transi tion regime} \\ \text{d} < 1.7\text{nm} & \text{molecular dynamics} \end{cases}$$

2.2.3. Collision Rate of Aerosol Particles

2.2.3.1. Molecular Dynamics

Collision rate of particles i and j per unit volume is given by molecular dynamics as

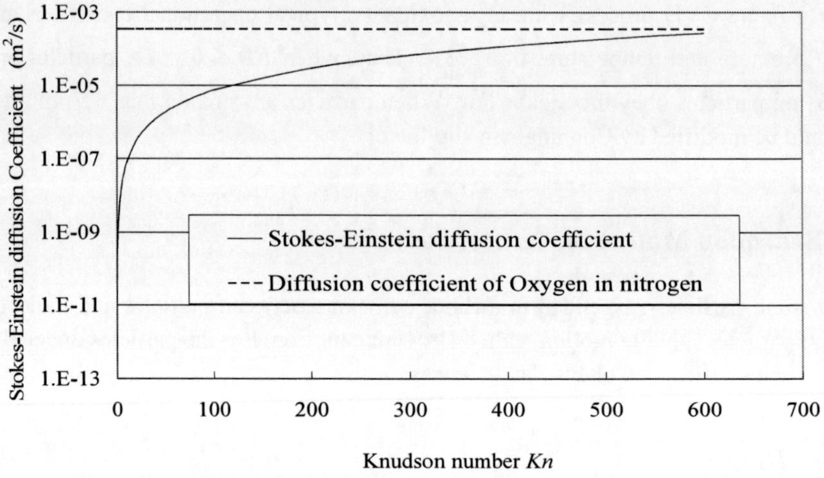

Figure 2.18. Stokes-Einstein diffusion coefficient *versus* Knudson number.

$$\dot{Z}_{ij} = \frac{1}{2} N_i N_j \sigma_{ij} \overline{v}_{ij} \qquad (2\text{-}42)$$

where N_i and N_j are number densities of particle i and particles j respectively, σ_{ij} is collision cross sectional area of particle i and particle j, and $\overline{v}_{ij}$ is the mean square root of the collision of particle i and particle j.

The collision cross sectional area of particle i and particle j is given by

$$\sigma_{ij} = \pi d_{ij}^2 \qquad (2\text{-}43)$$

where d_{ij} is the collision diameter of particle i and particle j, and is equal to

$$d_{ij} = \frac{1}{2}\left(d_i + d_j\right) \qquad (2\text{-}44)$$

as shown in Figure 2-19.

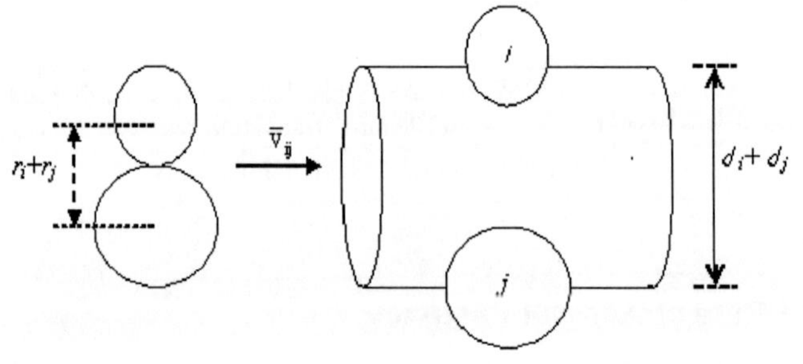

Figure 2.19. Illustration of collisions of particle i and particle j.

The mean square root of the collision of particle *i* and particle *j*., $\overline{v}_{ij}$, is

$$\overline{v}_{ij} = \sqrt{\overline{v}_i^2 + \overline{v}_j^2} \tag{2-45}$$

where $\overline{v}_i$ and $\overline{v}_j$ are mean square root of the speed of particle *i* and particle *j* and are given by

$$\overline{v}_i = \sqrt{\frac{8kT}{\pi m_i}} = \sqrt{\frac{48kT}{\pi^2 d_i^3 \rho_P}} \tag{2-46}$$

$$\overline{v}_j = \sqrt{\frac{8kT}{\pi m_j}} = \sqrt{\frac{48kT}{\pi^2 d_j^3 \rho_P}} \tag{2-47}$$

where *m* is the mass of particles.

Substituting equations (2-43) to (2-47) into equation (2-42), one can obtained the collision rate of particles *i* and *j* per unit volume that

$$\dot{Z}_{ij} = \sqrt{\frac{3kT}{4\rho_P}}\left(d_i + d_j\right)^2 \sqrt{\frac{1}{d_i^3} + \frac{1}{d_j^3}} N_i N_j \tag{2-48}$$

The distribution function of particle number in the size interval Δd can be written as [22]

$$y = \frac{\Delta N}{N \Delta d} = f(d) \tag{2-49}$$

where ΔN is the particle numbers of size *d* in the interval of Δd, *N* is the total particle numbers. From the definition of size distribution function, it meets that

$$\int_0^\infty y \Delta d = \int_0^\infty f(d) \Delta d = 1 \tag{2-50}$$

Thus, number densities of particles *i* and *j* are equal to

$$N_i = N \Delta d_i f(d_i) \tag{2-51}$$

$$N_j = N \Delta d_j f(d_j) \tag{2-52}$$

Substituting equations (2-51) and (2-52) into equation (2-48), it is obtained that

$$\dot{Z}_{ij} = N^2 \sqrt{\frac{3kT}{4\rho_P}} \left(d_i + d_j\right)^2 \sqrt{\frac{1}{d_i^3} + \frac{1}{d_j^3}} f(d_i) f(d_j) \Delta d_i \Delta d_j \tag{2-53}$$

The total collision rate per unit volume is then

$$\dot{Z} = N^2 \sqrt{\frac{3kT}{4\rho_P}} \int_0^\infty \int_0^\infty \left(d_i + d_j\right)^2 \sqrt{\frac{1}{d_i^3} + \frac{1}{d_j^3}} f(d_i) f(d_j) \Delta d_i \Delta d_j \tag{2-54}$$

It is written for simplicity as

$$\dot{Z} = \sqrt{\frac{3kT}{4\rho_P}} N^2 \Sigma_M \tag{2-55}$$

where the integral Σ_M is

$$\Sigma_M = \int_0^\infty \int_0^\infty \left(d_i + d_j\right)^2 \sqrt{\frac{1}{d_i^3} + \frac{1}{d_j^3}} f(d_i) f(d_j) \Delta d_i \Delta d_j \tag{2-56}$$

If particle size is represented by the normal distribution [22]

$$y = f(d) = \frac{1}{s\sqrt{2\pi}} e^{-\frac{(d-\bar{d})^2}{2s^2}} \tag{2-57}$$

where $\bar{d}$ is the mean particle size and s is the standard deviation. Since the size distribution function is a two parameters model, the integral Σ_M depends on both the mean size and the standard deviation. If the standard deviation is taken as

$$s = \bar{d}/\alpha \tag{2-58}$$

the integral Σ_M, after integration, can be expressed to

$$\Sigma_M = \frac{\chi_M \alpha^2}{2\pi} \sqrt{\bar{d}} \tag{2-59}$$

where χ_M is the integral

$$\chi_M = \int_0^\infty \int_0^\infty \frac{1}{\bar{d}^{2.5}} \left(d_i + d_j\right)^2 \sqrt{\frac{1}{d_i^3} + \frac{1}{d_j^3}} e^{-\frac{\alpha^2 (d_i - \bar{d})^2}{2\bar{d}^2}} e^{-\frac{\alpha^2 (d_i - \bar{d})^2}{2d^2}} \Delta d_i \Delta d_j \tag{2-60}$$

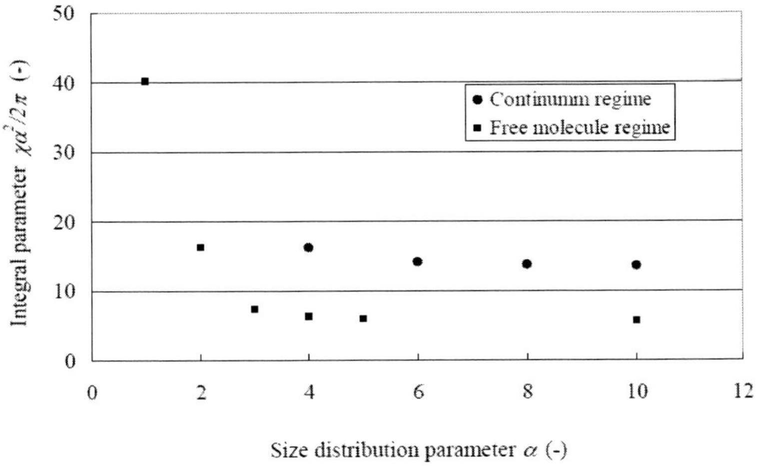

Figure 2.20. Integral parameter $\chi\alpha^2/2\pi$ *versus* size distribution parameter α.

Calculated results show that $\chi_M\alpha^2/2\pi$ decreases as α increases and is almost leveled off at a value of 5.353 when α is more than 3.0 as shown in Figure 2-20. The total collision rate per unit volume is then equal to

$$\dot{Z} = \sqrt{\frac{3kT}{4\rho_P}} N^2 \Sigma_M = 4.636\sqrt{\frac{kT\bar{d}}{\rho_P}} N^2 = 4.636\sqrt{\frac{kT\lambda}{\rho_P}}\sqrt{\frac{1}{Kn}} N^2 \tag{2-61}$$

2.2.3.2. Brownian-Smoluchowski Theory

Collisions between particles by Brownian motion are approached by a diffusion model as illustrated in Figure 2-21, for a static particle *i*, the diffusion rate or collision rate of particle *j* to particle *i* is given by [23]

$$\dot{Z}_{i\leftarrow j} = 4\pi D_P \left(r^2 \frac{\partial N}{\partial r} \right)_{r=(R_i+R_j)} = 4\pi D_{Pj} (R_i + R_j) N_j \tag{2-62}$$

where *r* is radial coordinator in the spherical coordination, R_i and R_j are the radii of particles *i* and *j*, respectively.

The collision rate of particles *i* and *j*, when particle *i* is moving, is then

$$\dot{Z}_{ij} = \frac{1}{2} 4\pi (D_{Pi} + D_{Pj})(R_i + R_j) N_i N_j \tag{2-63}$$

Substituting equations (2-41), (2-51) and (2-52) into equation (2-63), it is obtained that

$$\dot{Z}_{ij} = \frac{kT}{3\mu} N^2 \left(\frac{C_{C,i}}{d_i} + \frac{C_{C,j}}{d_j} \right) (d_i + d_j) f(d_i) f(d_j) \Delta d_i \Delta d_j \tag{2-64}$$

The total collision rate per unit volume is then

$$\dot{Z} = \frac{kT}{3\mu} N^2 \Sigma_B \tag{2-65}$$

where the integral Σ_B is

$$\Sigma_B = \int_0^\infty \int_0^\infty \left(\frac{C_{C,i}}{d_i} + \frac{C_{C,j}}{d_j} \right) (d_i + d_j) f(d_i) f(d_j) \Delta d_i \Delta d_j \tag{2-66}$$

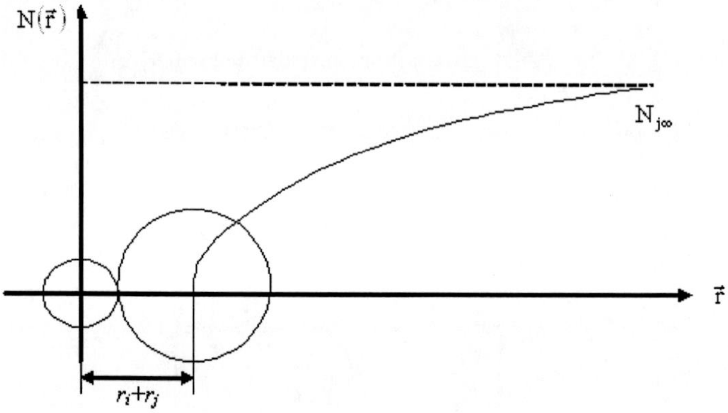

Figure 2.21. Illustration of particle collisions by Brownian motion.

Under the same particle size distribution as given by equation (2-57) and the size distribution parameter α defined by equation (2-58), the integral Σ_B is equal to

$$\Sigma_B = \frac{\lambda^2}{2\pi s^2} \int_0^\infty \int_0^\infty \left(C_{Ci} Kn_i + C_{Cj} Kn_j \right) \left(\frac{1}{Kn_i} + \frac{1}{Kn_j} \right) \times$$
$$e^{-\frac{\alpha^2 Kn^2}{2} \left(\frac{1}{Kn_i} - \frac{1}{Kn_j} \right)^2} e^{-\frac{\alpha^2 Kn^2}{2} \left(\frac{1}{Kn_j} - \frac{1}{Kn} \right)^2} \times \tag{2-67}$$
$$\frac{1}{Kn_i^2 Kn_j^2} \Delta Kn_i \Delta Kn_j$$

where Knudson number Kn, Kn_i and Kn_j are

$$Kn = \frac{\lambda}{\bar{d}} \tag{2-68}$$

$$\mathrm{Kn}_i = \frac{\lambda}{d_i}, \mathrm{Kn}_j = \frac{\lambda}{d_j} \tag{2-69}$$

The integral Σ_B is written for convenience as

$$\Sigma_B = \frac{\lambda^2 \chi_B}{2\pi s^2} = \frac{\chi_B \alpha^2}{2\pi} \mathrm{Kn}^2 \tag{2-70}$$

and the χ_B is the integral

$$\chi_B = \int_0^\infty \int_0^\infty \left(C_{Ci}\mathrm{Kn}_i + C_{Cj}\mathrm{Kn}_j\right)\left(\frac{1}{\mathrm{Kn}_i} + \frac{1}{\mathrm{Kn}_j}\right) \times$$
$$e^{-\frac{\alpha^2 \mathrm{Kn}^2}{2}\left(\frac{1}{\mathrm{Kn}_i} - \frac{1}{\mathrm{Kn}_j}\right)^2} e^{-\frac{\alpha^2 \mathrm{Kn}^2}{2}\left(\frac{1}{\mathrm{Kn}_j} - \frac{1}{\mathrm{Kn}}\right)^2} \times \tag{2-71}$$
$$\frac{1}{\mathrm{Kn}_i^2 \mathrm{Kn}_j^2} \Delta \mathrm{Kn}_i \Delta \mathrm{Kn}_j$$

Calculated results show that $\chi_B \alpha^2 / 2\pi$ decreases as α increases and is almost leveled off at a value of 14.0 when α is more than 4.0 as shown in Figure 2-20. The total collision rate per unit volume is then equal to

$$\dot{Z} = 14\,\mathrm{Kn}^2 \frac{kT}{3\mu} N^2 \tag{2-72}$$

2.2.4. Integral Collision Kernel

The decrease rate of particle/aggregate number density due to particle coalescence is then assumed to be

$$\dot{Z} = \beta N^2 \tag{2-73}$$

where $\dot{z}$ is particle/aggregate collision rate and β is an integral collision kernel which is independent of the particle number density N.

From equations (2-61) and (2-73), the integral collision kernel in the regime of molecular dynamics, β_M, is equal to

$$\beta_M = 4.636 \sqrt{\frac{kT\lambda}{\rho_P}} \sqrt{\frac{1}{\mathrm{Kn}}} \tag{2-74}$$

From equations (2-72) and (2-73), the integral collision kernel in the continuum regime, β_B, is equal to

$$\beta_B = 14\,Kn\,\frac{kT}{3\mu} \tag{2-75}$$

The integral collision kernel in the transition regime β_{TR} is given by

$$\beta_{TR} = \sqrt{\frac{kT\lambda}{\rho_P}}\,h(Kn) \tag{2-76}$$

where *h(Kn)* is a fitting function and is obtained by the interpolation between the two limiting regimes as

$$h(Kn) = \frac{\gamma\sinh(5Kn)}{1+\dfrac{\gamma\sinh(5Kn)}{p(Kn)}} \tag{2-77}$$

where sinh*(5Kn)* is the hyperbolic function, γ and *p(Kn)* are

$$\gamma = \sqrt{\frac{kT\rho_P}{\lambda\mu^2}} \tag{2-78}$$

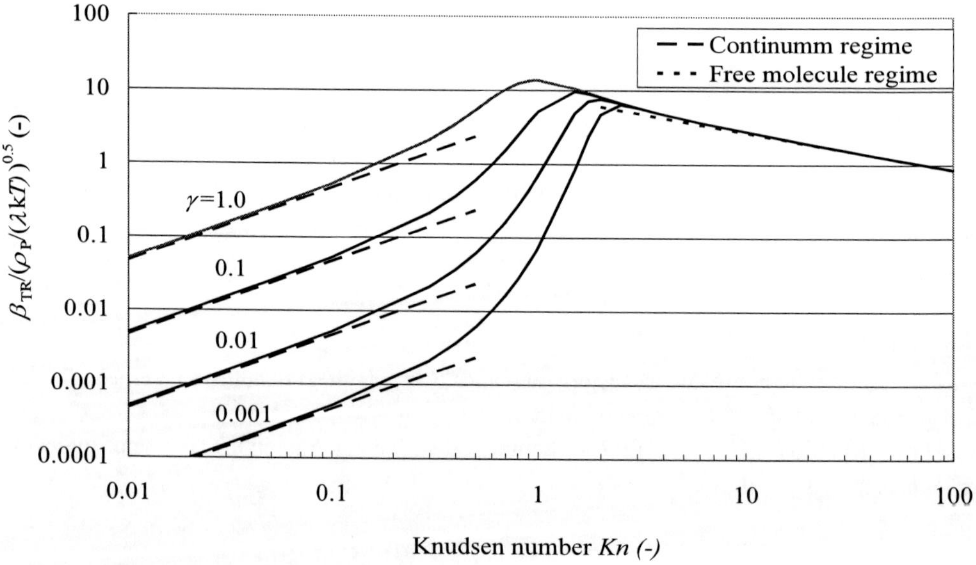

Figure 2.22. Integral collision kernel *versus* Knudsen number.

and

$$p(Kn) = \frac{8.495}{\sqrt{Kn}} + \frac{8.495}{Kn^2} \tag{2-79}$$

respectively.

The integral collision kernel $\beta_{TR}/\sqrt{\rho_P/kT\lambda}$ versus the Knudsen number is plotted in Figure 2-22 at different values of the parameter γ. It can be seen that when the Knudsen number is less than 0.1, the integral collision kernel β_{TR} approached to the integral collision kernel in the continuum regime; and when the Knudsen number is more than about 2, the integral collision kernel β_{TR} approached to the integral collision kernel in the free molecule regime.

2.2.5. Characteristic Collision Time of Nanoparticles

From Figure 2-22, it can be seen that Knudsen number is more than about 2, collisions between particles are in the free molecule regime. In flame CVD conditions, the mean free path of gas molecules is about 200nm as demonstrated in section 2.2.1, collisions between nanoparticles are then in the free molecule regime since when particles are less than 100nm, Knudsen number is more than about 2. From molecular dynamics, the characteristic collision time, τ_c, is calculated as

$$\tau_c = \frac{1}{N\sigma\overline{v}} \tag{2-80}$$

where collision cross sectional area and is equal to (Figure 2-19)

$$\sigma = \pi d_{ij}^2 = \pi\left(\frac{1}{2}(d_i + d_j)\right)^2 = \frac{\pi}{4}(d_i + d_j)^2 \tag{2-81}$$

The mean square root of the speed of particle i and particle j and are given by equations (2-46) and (2-47) and are given here repeatedly as

$$\overline{v}_i = \sqrt{\frac{48kT}{\pi^2 d_i^3 \rho_P}} \tag{2-82}$$

$$\overline{v}_j = \sqrt{\frac{48kT}{\pi^2 d_j^3 \rho_P}} \tag{2-83}$$

The mean square root of the collision of particle i and particle j., $\overline{v}_{ij}$, is then equal to

$$\overline{V}_{ij} = \sqrt{\overline{V}_i^2 + \overline{V}_j^2} = \sqrt{\frac{48kT}{\pi^2 \rho_P}} \sqrt{\frac{1}{d_i^3} + \frac{1}{d_j^3}} \tag{2-84}$$

The characteristic collision time, τ_c, is then equal to

$$\tau_c = \frac{1}{N\sigma\overline{v}} = \frac{1}{N}\sqrt{\frac{\rho_P}{3kT}}\sqrt{\frac{d_i^3 d_j^3}{(d_i + d_j)^4 (d_i^3 + d_j^3)}} \tag{2-85}$$

For equal sized nanoparticles, the characteristic collision time, τ_c, is reduced to

$$\tau_c = \frac{1}{N\sigma\overline{v}} = \frac{1}{N}\sqrt{\frac{\rho_P}{96kTd}} \tag{2-86}$$

For particle density ρ_P=3800kg/m^3 and temperature T=1273K, the characteristic collision time, τ_c, calculated by equation (2-86) is shown in Figure 2-23, it could be seen that the characteristic collision time for titania nanoparticles under flame CVD conditions ranges from about 10μs to 1s.

In the calculations in Figure 2-23, the particle number density depends on particle mass fraction. For particles uniformly suspended in a fluid, the particle number density, N, equals

$$N = \frac{\rho_f f_P}{\frac{\pi}{6} d^3 \rho_P} \tag{2-87}$$

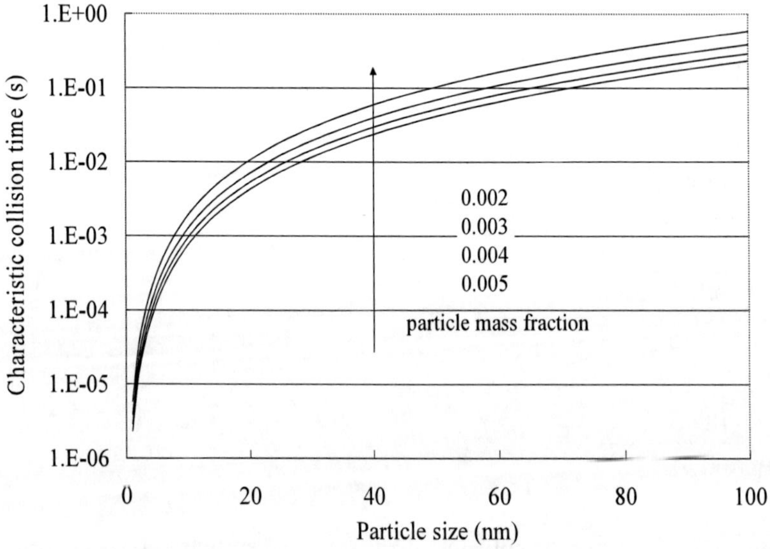

Figure 2.23. Characteristic collision time.

where ρ_f is the density of the fluid and f_P is particle mass fraction suspended in the fluid. For titania particles (ρ_P=3800kg/m^3) uniformly suspended in air at temperature of 1273K, the calculated particle number density, *N*, is shown in Figure 2-24. One can see that the particle number density increases as particle size decreases and particle mass fraction increases. For particles about 1nm, the particle number density ranges about 10^{20} to 10^{21} m^{-3}, which is under typical operational conditions of flame CVD processes as given in Tables 1-1 and 1-3.

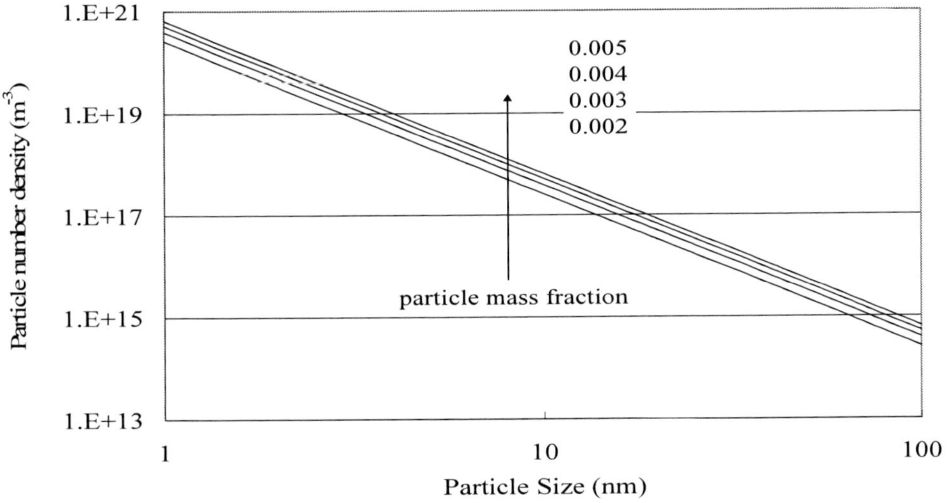

Figure 2.24. Particle number density *versus* particle size and particle mass fraction.

Comparisons between characteristic collision time and coalescence time are plotted in Figure 2-25 under typical operational conditions of flame CVD processes. It shows that the characteristic coalescence time is about one order shorter than the characteristic collision time. Considering the effect of coalescence energy release, particle temperature can be several hundred degrees higher than the gas temperature, and thus the coalescence time can be significantly reduced by orders of magnitude [5]. This is consistent with TEM images showing the particles are predominantly spherical.

2.3. Growth of Aerosol Particles in Aerosol Processes

The development of the size distribution by coagulation can be described by the coagulation equation proposed by Muller [24], which is a nonlinear integral differential equation. Although a number of methods e.g., sectional technique [25-27], self-preserving model [28-30], has been developed to solve the coagulation equation numerically. The main disadvantage of Muller's method is that it is difficult to integrate it with the mass, momentum and energy equations of gas domain for process design, which is encountered with aerosol processes.

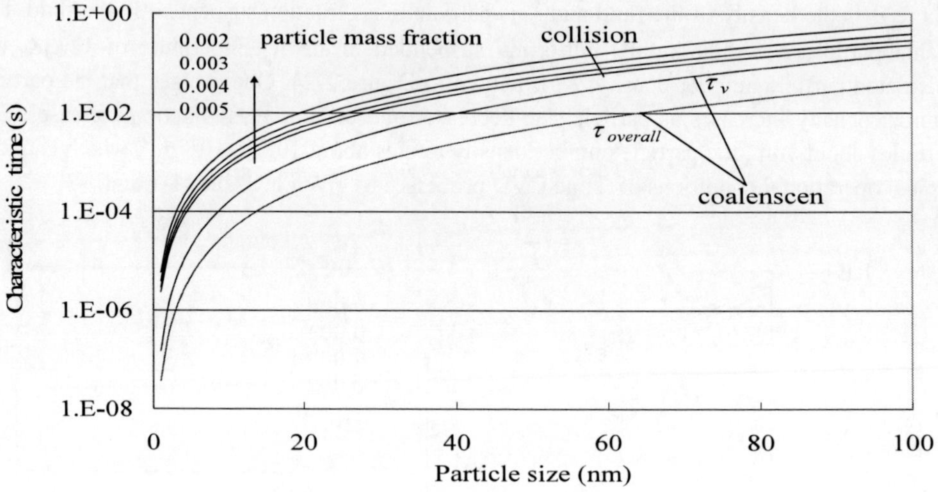

Figure 2.25. Comparisons of characteristic collision and coalescence times.

From the viewpoint of the balance of particle number density, the development of particle size by coagulation in aerosol processes could be treated by the principle of fluid dynamics with mass balance equations.

2.3.1. Balance Equations of Particle Numbers

Let us consider one dimensional flame CVD process as illustrated in Figure 2-26, where *A* is the cross sectional area, *u* is macroscopic velocity, the number of particles enter the element control volume *Adx* per unit time, $\dot{n}_{in}$, is contributed by convection flow of particles along with the gas domain and particle's diffusion within the gas domain and is equal to

$$\dot{n}_{in} = uAN - D_p A \frac{\partial N}{\partial x} \tag{2-88}$$

and the number of particles exit the element control volume *Adx* per unit time, $\dot{n}_{out}$, is

$$\dot{n}_{out} = uAN + uA\frac{dN}{dx}dx - D_p A\frac{\partial N}{\partial x} - D_p A\frac{d^2N}{dx^2}dx \tag{2-89}$$

Assuming that particle numbers increased by chemical reactions per unit time and per unit volume is $\dot{n}_r$ and particle numbers decreased by particle coalescence per unit time and per unit volume is $\dot{n}_c$, the balance equation of particle numbers is

$$Adx\frac{dn}{dt} = \dot{n}_{in} - \dot{n}_{out} + Adx\,\dot{n}_r - Adx\,\dot{n}_c \tag{2-90}$$

Substituting equations (2-88) and (2-89) into (2-90), the balance equation of particle numbers is obtained that

$$A dx \frac{dn}{dt} = uAN - D_P \frac{dN}{dx} - uAN - uA\frac{dN}{dx}dx + D_P A\frac{dN}{dx} + D_P A\frac{d^2N}{dx^2}dx + Adx\dot{n}_r - Adx\dot{n}_c \quad (2\text{-}91)$$

the balance equation of particle numbers is finally

$$\frac{dn}{dt} = -u\frac{dN}{dx} + D_P\frac{d^2N}{dx^2} + \dot{n}_r - \dot{n}_c \quad (2\text{-}92)$$

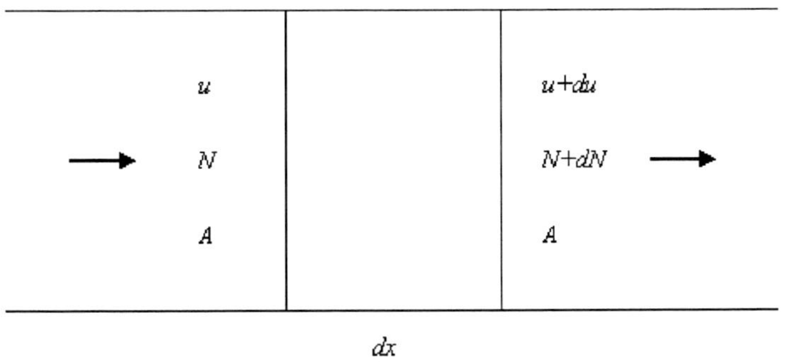

Figure 2.26. Illustration of one dimensional flame CVD process.

2.3.2. One Dimensional Steady Flow

The particle numbers decreased by particle coalescence per unit time and per unit volume, $\dot{n}_c$, is equal to

$$\dot{n}_c = \dot{Z} \quad (2\text{-}93)$$

Substituting equation (2-73) into equation (2-93), the particle numbers decreased by particle coalescence per unit time and per unit volume, $\dot{n}_c$, is

$$\dot{n}_c = \dot{Z} = \beta N^2 \quad (2\text{-}94)$$

Particle mass increased by chemical reactions per unit time and per unit volume, $\dot{W}_P$ is equal to

$$\dot{W}_P = \dot{r}_v M_P = \kappa k_v C^{\varepsilon} M_P \quad (2\text{-}95)$$

where $\dot{r}_v$ is the chemical reaction rate per unit volume, k_v is reaction rate constant per unit volume, C is the concentration of a reactant, ε is the power of the order of the reactions, κ is the stoichiometric ratio and M_P is the mole weight of the particles.

The particle numbers increased by chemical reactions per unit time and per unit volume, $\dot{n}_r$, is then equal to

$$\dot{n}_r = \frac{\dot{W}_P}{\frac{\pi}{6} d_0^3 \rho_P} = \frac{\kappa k_v C^{\varepsilon} M_P}{\frac{\pi}{6} d_0^3 \rho_P} \tag{2-96}$$

where d_0 is the size of monomers of nanoparticles, ρ_P is the density of nanoparticles.

For steady flow and neglecting the contribution of Brownian motion, the balance equation of particle numbers becomes

$$0 = -u\frac{dN}{dx} + \dot{n}_r - \dot{n}_c \tag{2-97}$$

Since flame CVD process is a high temperature process, one can assume that the product is generated by gas phase oxidation reactions of precursors completely and instantaneously at the initial plane, then the balance equation of particle numbers of one-dimensional steady flow is reduced to

$$0 = -u\frac{dN}{dx} - \dot{n}_c \tag{2-98}$$

Substituting equation (2-94) into to equation (2-98), the balance equation of particle numbers of one-dimensional steady flow is further simplified to

$$-u\frac{dN}{dx} = \beta N^2 \tag{2-99}$$

By integrating equation (2-99), one obtains that

$$\frac{N}{N_0} = \frac{1}{1 + \frac{N_0}{u}\int_0^x \beta dx} \tag{2-100}$$

Let

$$\xi = \frac{x}{L} \tag{2-101}$$

where L is the reactor or flame length, equation (2-100) becomes

$$\frac{N}{N_0} = \frac{1}{1 + \frac{LN_0}{u} \int_0^{\xi} \beta d\xi} \tag{2-102}$$

At the reactor outlet or the exit of flame, equation (2-102) becomes

$$\frac{N}{N_0} = \frac{1}{1 + \frac{LN_0}{u} \int_0^{1} \beta d\xi} \tag{2-103}$$

Equation (2-103) can be written as

$$\frac{N}{N_0} = \frac{1}{1 + N_{grw}} \tag{2-104}$$

where N_{grw} is particle growth number and equals

$$N_{grw} = \frac{\tau}{\tau_{grw}} \tag{2-105}$$

where τ is the residence time

$$\tau = \frac{L}{u} \tag{2-106}$$

and τ_{grw} is characteristic particle growth time

$$\tau_{grw} = \frac{1}{N_0 \int_0^{1} \beta d\xi} \tag{2-107}$$

For compounds produced by reactions in the gas phase, which subsequently grow into aggregates by particle coalescence, the mass balance of the particles can be written as

$$n\rho_P \frac{\pi}{6} \overline{d}^3 = \text{constant} \tag{2-108}$$

Combing equations (2-104) and (2-108), one can obtaine that

$$\frac{\bar{d}}{d_0} = (1 + N_{grw})^{1/3} \tag{2-109}$$

Assuming that the product is generated by gas phase oxidation reactions of precursors completely and instantaneously at the initial plane, the particle numbers increased by chemical reactions per unit time and per unit volume, $\dot{n}_r$, is then equal to

$$\dot{n}_r = \frac{\dot{W}_P}{\frac{\pi}{6} d_0^3 \rho_P} = \frac{\kappa \dot{q}_{m,pre} M_P}{\frac{\pi}{6} d_0^3 \rho_P M_{pre}} \tag{2-110}$$

where $\dot{q}_{m,pre}$ is the mass flow rate of precursors, and M_{pre} is the molecular weight of precursor.

Assuming that the size of monomers of nanoparticles is 1nm, the particle number density by the chemical reactions, N_0, is then

$$N_0 = \frac{\dot{n}_r}{q_{v,gas}} = \frac{6}{\pi} 10^{27} \varphi \frac{q_{m,precur} M_P}{q_{v,gas} \rho_P M_{precur}} \tag{2-111}$$

where $\dot{q}_{v,gas}$ is the total gas volume flow rate in reactor. Equation (2-109) then becomes

$$\bar{d} = (1 + N_{grw})^{1/3} \tag{2-112}$$

where the mean size $\bar{d}$ is in the unit of nanometers.

It has been shown, in Chapter 1, that the mean size $\bar{d}$ could be fitted with equation (1-1), which is similar to equation (2-112), though the particle growth number has slightly different from that defined by equations (2-105) to (2-107).

Combining equations (2-76) to (2-79), (2-105) to (2-107), and (2-111) to (2-112), the mean particle size can be calculated at given gas and particle properties, and operational conditions. The integral collision kernel can then be obtained from equations (2-74), (2-75), and (2-76) to (2-79) for the free molecule, continuum, and transition regime, respectively. Comparisons on the collision kernel could thus be made between such obtained values and those predicted by equations (2-76) to (2-79) with the measured Knudsen number.

Titania and silica nanoparticles are synthesized by flame CVD process with CO and propane laminar diffusion flames, and the precursors of $TiCl_4$ and $SiCl_4$ for titania and silica, respectively, which are introduced into the burner by dry nitrogen or air through a bubbler. Details about the experiments are described in Chapter 1, the experimental conditions and experimental results such as gas and precursor flow rates, flame temperature and particle size are given in Table 2-1. Predicted particle size of the experimental runs is also given in Table 2-1. Grossly the predicted particle size agrees reasonably well with experimental results.

Table 2.1. Experimental conditions and experimental results

Run No.[a]	Flame temperature (K)	Visual flame length (m)	Flow rate ($mmol\cdot s^{-1}$) C_3H_8/CO	Flow rate ($mmol\cdot s^{-1}$) Air/O_2 (oxidant)	Flow rate ($mmol\cdot s^{-1}$) Air/N_2 (carrier gas)	$TiCl_4/SiCl_4$ Flow rate ($g\cdot min^{-1}$)	Particle size[d] measured (nm)	Particle size predicted (nm)	Mean free path of gas molecules (nm)
1	1287	0.025	2.031	3.616	5.952[b]	0.12	50	52.8	308
2	1431	0.02	2.240	3.757	4.658[c]	0.15	70	69.9	343
3	1300	0.02	0.217	8.929	0.372	0.069	40	42.4	311
4	1250	0.04	0.129	5.539	0.749	0.0686	24.3	48.6	280
5	1500	0.08	0.259	5.988	1.123	0.113	16.6	50.7	336
6	1440	0.08	0.259	5.988	1.497	0.135	32.3	55.7	323
7	1500	0.06	0.259	6.737	0.374	0.0703	42.9	35.8	336
8	1225	0.04	0.259	8.234	0.374	0.0965	23.6	36.2	275
9	1173	0.035	0.259	9.880	0.374	0.0965	23.5	32.8	263
10	1150	0.04	0.259	11.078	0.374	0.079	16.6	31.0	258
11	1005	0.04	0.259	11.078	1.123	0.113	24.8	38.8	225
12	980	0.04	0.259	10.78	1.497	0.135	26.0	42.7	220
13	1120	0.04	0.259	11.976	0.374	0.079	16.5	30.6	251
14	1753	0.20	0.388	4.491	0.374	0.07	61.9	30.6	393
15	1708	0.175	0.388	5.614	0.374	0.06	56.7	50.9	383
16	1695	0.13	0.388	6.737	0.374	0.06	56.5	42.7	380
17	1640	0.10	0.388	7.859	0.374	0.06	48.9	35.7	368
18	1623	0.09	0.388	8.982	0.374	0.06	37.7	31.0	364
19	1600	0.09	0.388	8.982	0.749	0.076	39.3	28.5	359

[a]: Run No.1 and 2 are TiO_2 and SiO_2 from CO/O_2 laminar diffusion flame respectively, No.3 is TiO_2 from C_3H_8/O_2 laminar diffusion flame, Run No.4-19 are TiO_2 from C_3H_8/air flame; [b]: 0.372mmol/s is used as carrier gas, the rest is mixed with O_2; [c]: 0.119mmol/s is used as carrier gas, the rest is mixed with O_2; Run No. 2 and 3 were determined by specific surface area ($37m^2/g$ for SiO_2 and $39m^2/g$ for TiO_2), the rest are TEM sizes.

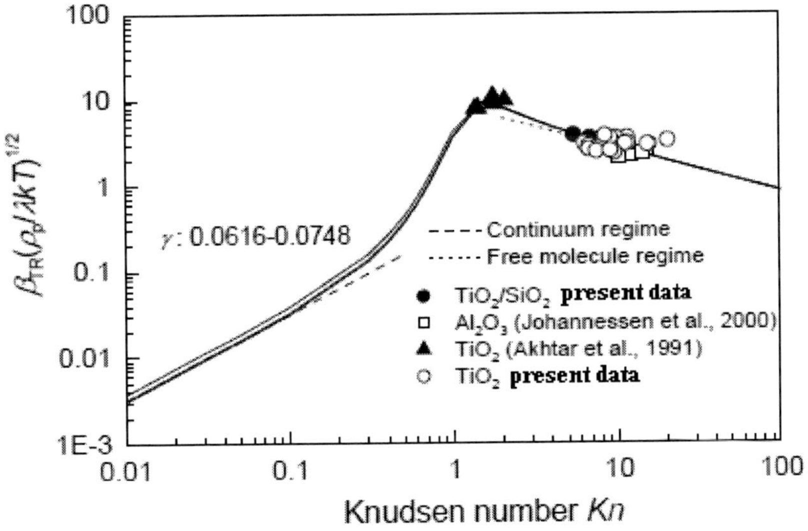

Figure 2.27. Comparisons on the integral collision kernel between predicted and experimental values.

Alumina nanoparticles synthesized by flame CVD process [4,31] are also compared for C1-C6 methane diffusion flames, the flame length and flame temperature are taken as 0.12m and 1500K, respectively, for the six flames as measured by the authors, the particles density is taken as 4000kg/m^3.

Titania submicron particles with mean size ranging from 0.18 to 0.27μm synthesized in a tube furnace reactor [32] are compared. The effective reactor length is taken the isothermal reaction zone, which is about 0.5m at the reactor temperature of 1200K and 0.7m at the reactor temperature between 1400 and 1723K as measured by the authors.

Comparison on the integral collision kernel β versus the particle Knudsen number Kn is shown in Figure 2-27. It can be seen that the collision kernel predicted by Equations (2-76) to (2-79) agrees well with the experimental results in the Knudsen number Kn range from about 1.5 to 20.

Nomenclature

A cross sectional area, m^2
C concentration, kmol m^{-3}
C_C Cunning slip factor
d particle size, m
$\bar{d}$ mean particle size, m
Δd size interval, m
D_P Stokes-Einstein diffusion coefficient, m^{2}s^{-1}
F_D drag force, N
f_P mass fraction of particle suspended in a fluid
j particle number flux, m^{-2}s^{-1}
k Boltzmann constant, 1.3805*10^{-23}JK^{-1}
k_v reaction rate constant per unit volume, kmol m^{-3}
Kn Knudsen number ($=\lambda/d$)
L reactor or flame length, m
m mass of particles, kg
M_P molecular weight of particles, kg kmol^{-1}
M_{pre} molecular weight of precursors, kg kmol^{-1}
N molecule or particle number density, m^{-3}
ΔN particle number density of size d in the interval Δd
n particle numbers
$\dot{n}_c$ decrease rate of particle numbers due to particle coalescence, m^{-3}s^{-1}
$\dot{n}_r$ decrease rate of particle numbers due to chemical reaction, m^{-3}s^{-1}
N_{grw} particle growth number defined by equation (2-105)
$\dot{q}_{v,gas}$ gas volume flow rate, m^{3}s^{-1}
$\dot{q}_{m,pre}$ mass flow rate of precursors, kgs^{-1}
r coordinate in spherical coordination, m
$\dot{r}_v$ chemical reaction rate per unit volume, kmol s^{-1}
R radius of particles, m
s standard deviation of size distribution, m

T temperature, K
u macroscopic stream velocity, m/s
$\bar{v}$ mean square root of particle speed, ms^{-1}
$\bar{v}_{ij}$ mean square root of coalescing particles, ms^{-1}
W_P particle mass increased by chemical reaction, kgs^{-1}
x distance from chemical reaction plane, m
y particle number frequency per micron, m^{-1}
$\dot{z}$ particle collision rate, $m^{-3}s^{-1}$
$\dot{Z}_{ij}$ particle collision rate of particles i and j, $m^{-3}s^{-1}$

Greek Letters

α particle size distribution parameter defined by equation (2-58)
β integral collision kernel, m^3s^{-1}
χ_M integral defined by equation (2-60)
χ_B integral defined by equation (2-71)
ε power of reaction order
κ stoichiometry ratio
μ fluid viscosity, Pas
λ mean free path of gas molecule, m
σ collision cross sectional area, m^2
ρ_f fluid density, kgm^{-3}
ρ_P particle density, kgm^{-3}
τ residence time, s
τ_c collision time, s
τ_{grw} particle growth time, s
υ_0 molecular size, m
ξ x/L
Σ_B integral in equation (2-66)
Σ_M integral in equation (2-56), $m^{0.5}$

Subscripts

0: position $x=0$
B continuum regime
M free molecule regime
i i sized particles
j j sized particles
in inter an element control volume
out exit an element control volume
TR transition regime

REFERENCES

[1] Frenkel J. (1945). The viscous flow in crystal bodies. *J. Phys.*, 9, 385.

[2] Koch W., Friedlander S.K. (1990). The effect of particle coalescence on the surface area of a coagulating aerosol. *J. Colloid and Interface Sci.*, 140, 419.

[3] Kruis F. E., Kusters K. A., Pratsinis S. E., Scarlett B. (1993). A simple model for the characteristics of aggregate particles undergoing coagulation and sintering. *Aerosol Sci. Technol.*, 19, 514.

[4] Johannessen T., Pratsinis S. E., Livbjerg H. (2000). Computational fluid-particle dynamics for the flame synthesis of alumina particles. *Chem. Eng. Sci.*, 55, 177.

[5] Lehtinen K. E. J, Zachariah M. R. (2001). Effect of coalescence energy release on the temporal shape evolution of nanoparticles. *Physical Review B*, 63, 205402-1-7.

[6] Hiram Y., Nir A. (1983). A simulation of surface tension driven coalescence. *J. of Colloid and Interface Sci.*, 95, 462.

[7] Friedlander S.K., Wu M.K. (1994). Linear rate law for the decay of the excess surface area of a coalescing solid particle. *Physical Review B*, 49, 3622.

[8] Yadha V., Helble J.J. (2004). Modeling the coalescence of heterogenous amorphous particles. *J. Aerosol Sci.*, 35, 665.

[9] Garabedian R.S., Helble J.J. (2001). A Model for the viscous coalescence of amorphous particles. *J. Colloid and Interface Sci.*, 234, 248.

[10] Cho K., Biswas P. (2006). A geometrical sintering model (GSM) to predict surface area change, *J. Aerosol Sci.*, 37, 1378.

[11] Nichols F.A., Mullins W.W. (1965). Morphological changes of a surface of revolution due to capillarity-induced surface diffusion. *J. Appl. Phys.*, 36, 1826.

[12] Guo S.J. (1998). *Theory of Powder Sintering*, Metallurgy Industry Press: Beijing (in Chinese).

[13] Kuczynski G.C. (1949). Self-diffusion in sintering of metallic particles. *Trans. AIME*, 185, 169.

[14] Kobata A., Kusakabe K., Morroka S. (1991). Growth and transformation of TiO_2 crystallites in aerosol reactor, *AIChE J.*, 37, 347.

[15] Seto, T., Shimada, M., and Okuyama, K. (1995). Evaluation of sintering of nanometer-sized titania using aerosol method, Aerosol Sci. and Technol., 23, 183.

[16] Astier, M., and Vergnon, P. (1976). Determination of the diffusion coefficients from sintering data of ultrafine oxide particles. *J. Solid State Chemistry,* 19, 67.

[17] German, R.M., (1996). *Sintering theory and practice*, New York: Plenum Press.

[18] Rahaman, M.N., (1995). *Ceramic processing and sintering*, New York: Marcel Dekker Inc.

[19] Hebrard, J.L., Nortier, P., Pijolat, M., and Soustelle, M. (1990). Initial sintering of submicrometer titania anatase powder, *J. American Ceramic Society*, 73, 79.

[20] Hidy G.M., and Brock J.R. (1965). Some remarks about the coagulation of aerosol particles by Brownian motion. *J. of Colloid Science, 20,* 477.

[21] Otto E., and Fissan H. (1999). Brownian coagulation of submicron particles. *Adv. Powder. Technol.*, 10, 1.

[22] Allen T. (1990). *Particle Size Measurement*, 4th edn., London: Chapman and Hall.

[23] Fuchs N. A. (1964). *Mechanics of Aerosols.* New York: Pergamon Press.

[24] Muller H. (1928). Zur allgemeinen theorie der raschen koagulation. *Kolloidbeihefte*, 27, 223.

[25] Gelbard F. (1990). Modeling multicomponent aerosol particle growth by vapor Condensation. *Aerosol Sci. Technol.*, 12, 399.

[26] Wu J.J., Flagan R.C. (1988). A discrete-sectional solution to the aerosol dynamics equation. *J. Colloid Interface Sci.*, 123, 339.

[27] Landgrebe J.D., Pratsinis S.E. (1990). A discrete-sectional model for powder production by gas phase chemical reaction and aerosol coagulation in the free-molecular regime. *J. Colloid Interface Sci.,* 139, 63.

[28] Friedlander S.K., Wang C.S. (1966). The self-preserving particle size distribution for coagulation by Brownian motion. *J. Colloid Interface Sci.*, 22, 126.

[29] Lai F.S., Friedlander S.K., Pich J., et al. (1972). The self-preserving particle size distribution for coagulation in the free-molecule regime. *J. Colloid Interface Sci.*, 39, 395.

[30] Graham S.C., Robinson A. (1976). A comparison of numerical solutions to the self-preserving particle size distribution for aerosol coagulation in the transition regime. *J. Aerosol Sci.*, 7, 261.

[31] Kruis F.E., Kusters K.A., Pratsinis S.E., and Scarlett B. (1993). A simple model for the characteristics of aggregate particles undergoing coagulation and sintering. *Aerosol Science and Technology*, 19, 514.

[32] Akhtar M.K., Xiong Y., and Pratsinis S.E. (1991). Vapor synthesis of titania powder by titanium tetrachloride oxidation. *AIChE J.*, 37, 1561.

Chapter 3

SIMULATION OF PARTICLE GROWTH IN FLAME CVD PROCESS BY COMPUTATIONAL FLUID DYNAMICS

3.1. INTRODUCTION OF COMPUTATIONAL FLUID DYNAMICS [1]

3.1.1. Physical Models

3.1.1.1. Mass Balance Equation

For 2D axisymmetric geometries steady flow, the continuity equation is given by

$$\frac{\partial}{\partial x}(\rho u)+\frac{\partial}{\partial r}(\rho v)+\frac{\rho v}{r}=S_m \tag{3-1}$$

where x is the axial coordinate, r is the radial coordinate, u is the axial velocity, and v is the radial velocity. The source Sm is the mass added to the continuous phase from the dispersed second phase (e.g., due to vaporization of liquid droplets) and any other sources.

3.1.1.2. Momentum Conservation Equations

For 2D axisymmetric geometries steady flow, the axial and radial momentum conservation equations are given by

$$\begin{aligned}&\frac{1}{r}\frac{\partial}{\partial x}(r\rho uu)+\frac{1}{r}\frac{\partial}{\partial r}(r\rho vu)=\\&-\frac{\partial p}{\partial x}+\frac{1}{r}\frac{\partial}{\partial x}\left[r\mu\left(2\frac{\partial u}{\partial x}-\frac{2}{3}(\nabla\cdot\vec{v})\right)\right]+\frac{1}{r}\frac{\partial}{\partial r}\left[r\mu\left(2\frac{\partial u}{\partial r}+\frac{\partial v}{\partial x}\right)\right]+F_x\end{aligned} \tag{3-2}$$

$$\begin{aligned}&\frac{1}{r}\frac{\partial}{\partial x}(r\rho uv)+\frac{1}{r}\frac{\partial}{\partial r}(r\rho vv)=-\frac{\partial p}{\partial r}+\frac{1}{r}\frac{\partial}{\partial x}\left[r\mu\left(\frac{\partial v}{\partial x}+\frac{\partial u}{\partial r}\right)\right]\\&+\frac{1}{r}\frac{\partial}{\partial r}\left[r\mu\left(2\frac{\partial v}{\partial r}-\frac{2}{3}(\nabla\cdot\vec{v})\right)\right]-2\mu\frac{v}{r^2}+\frac{2}{3}\frac{\mu}{r}(\nabla\cdot\vec{v})+\rho\frac{w^2}{r}+F_r\end{aligned} \tag{3-3}$$

where

$$\nabla \cdot \vec{v} = \frac{\partial u}{\partial x} + \frac{\partial v}{\partial r} + \frac{v}{r} \tag{3-4}$$

and w is the swirl velocity, μ is gas viscosity, p is static pressure, F_x and F_r are the axial and radial components of external body force.

3.1.1.3. Energy Conservation Equation

Conservation of energy is described by

$$\frac{\partial}{\partial t}(\rho E) + \nabla \bullet \left(\vec{v}(\rho E + p)\right) = \nabla \bullet \left(k_{eff} \nabla T - \sum h_j \vec{J}_j + \left(\bar{\bar{\tau}}_{eff} \bullet \vec{v}\right)\right) + S_h \tag{3-5}$$

where k_{eff} is the effective conductivity ($k+k_t$, where k_t is the turbulent thermal conductivity, defined according to the turbulence model), and $\vec{J}_j$ is the diffusion flux of species j. The first three terms on the right-hand side of Equation (3-5) represent energy transfer due to conduction, species diffusion, and viscous dissipation, respectively. S_h includes the heat of chemical reaction, and any other volumetric heat sources.

In equation (3-5)

$$E = h - \frac{p}{\rho} + \frac{v^2}{2} \tag{3-6}$$

where sensible enthalpy h is defined for ideal gases as

$$h = \sum_j Y_j h_j \tag{3-7}$$

and for incompressible flows as

$$h = \sum_j Y_j h_j + \frac{p}{\rho} \tag{3-8}$$

where Y_j is the mass fraction of species j and

$$h_j = \int_{T_{ref}}^{T} c_{p,j} dT \tag{3-9}$$

where T_{ref} is reference temperature and is taken as 298.15 K.

3.1.1.4. Species Transport Equations

To solve conservation equations for chemical species, the conservation equation for local mass fraction of each species, Y_i, takes the following general form

$$\nabla \cdot \left(\rho \vec{v} Y_i \right) = -\nabla \vec{J}_i + R_i + S_i \tag{3-10}$$

where R_i is the net rate of production of species i by chemical reaction and S_i is the rate of creation by addition from the dispersed phase plus any other sources.

The equation will be solved for N-1 species where N is the total number of fluid phase chemical species present in the system. Since the mass fraction of the species must sum to unity, the N_{th} mass fraction is determined as one minus the sum of the N-1 solved mass fractions. To minimize numerical error, the N_{th} species should be selected as that species with the overall largest mass fraction, such as N_2 when the oxidizer is air.

In equation (3-10), $\vec{J}_i$ is the diffusion flux of species i, which arises due to concentration gradients, and can be written as

$$\vec{J}_i = -\left(\rho D_{i,m} + \frac{\mu_t}{Sc_t} \right) \nabla Y_i \tag{3-11}$$

where Sc_t is the turbulent Schmidt number ($\mu_t/\rho D_t$), where μ_t is the turbulent viscosity and D_t is the turbulent diffusivity.

3.1.2. Turbulent Model

3.1.2.1. Reynolds Stress Transport Equations

The Reynolds stress model (RSM) closes the Reynolds-averaged Navier-Stokes equations by solving transport equations for the Reynolds stresses, together with an equation for the dissipation rate. This means that five additional transport equations are required in 2D flows. The transport equations for the transport of the Reynolds stresses may be written as follows

$$\begin{aligned} \frac{\partial}{\partial t}\left(\rho \overline{u_i' u_j'}\right) + \frac{\partial}{\partial x_k}\left(\rho u_k \overline{u_i' u_j'}\right) &= -\frac{\partial}{\partial x_k}\left[\rho \overline{u_i' u_j' u_k'} + \overline{p\left(\delta_{kj} u_i' + \delta_{ik} u_j'\right)} \right] \\ &+ \frac{\partial}{\partial x_k}\left(\mu \frac{\partial}{\partial x_k} \overline{u_i' u_j'} \right) - \rho\left(\overline{u_i' u_k'} \frac{\partial u_j}{\partial x_k} + \overline{u_j' u_k'} \frac{\partial u_i}{\partial x_k} \right) - \rho\beta\left(g_i \overline{u_j' \theta} + g_j \overline{u_i' \theta} \right) \\ &+ \overline{p\left(\frac{\partial u_i'}{\partial x_j} + \frac{\partial u_j'}{\partial x_i} \right)} - 2\mu \overline{\frac{\partial u_i'}{\partial x_k} \frac{\partial u_j'}{\partial x_k}} - 2\rho\Omega_k \left(\overline{u_j' u_m'} \varepsilon_{ikm} + \overline{u_i' u_m'} \varepsilon_{jkm} \right) \end{aligned} \tag{3-12}$$

The first term in the left is local derivative, and the second term is convection, C_{ij}. The first term in the right is turbulent diffusion, $D_{T,ij}$, the second term is molecular diffusion, $D_{L,ij}$, the third is stress production, P_{ij}, the fourth is buoyancy production, G_{ij}, the fifth is pressure strain, ϕ_{ij}, the last but one is dispassion, ε_{ij}, the last is the production by system rotation, F_{ij}, which are needed to close the equations.

The turbulent diffusion, $D_{T,ij}$, can be given as

$$D_{T,ij} = \frac{\partial}{\partial x_k}\left(\frac{\mu_t}{\sigma_k}\frac{\partial \overline{u_i'u_j'}}{\partial x_k}\right) \tag{3-13}$$

where σ_k is 0.82 in the generalized gradient-diffusion model, and is 1.0 in the standard and realizable k-ε model. The turbulent viscosity, μ_t, is calculated by

$$\mu_t = \rho C_\mu \frac{k^2}{\varepsilon} \tag{3-14}$$

where $C_\mu = 0.09$, and the turbulent kinetic energy is obtained by taking the trace of the Reynolds stress tensor $k = \frac{1}{2}\overline{u_i'u_i'}$ and is modeled near wall zones by

$$\frac{\partial}{\partial t}(\rho k)+\frac{\partial}{\partial x_i}(\rho k u_i) = \frac{\partial}{\partial x_j}\left((\mu+\frac{\mu_t}{\sigma_k})\frac{\partial k}{\partial x_j}\right)+\frac{1}{2}(P_{ii}+G_{ii}) - \rho\varepsilon\left(1+2M_t^2\right) \tag{3-15}$$

where M_t is turbulent Mach number and is defined by

$$M_t = \sqrt{\frac{k}{a^2}} \tag{3-16}$$

where a is sound speed and is equal to

$$a \equiv \sqrt{\gamma RT} \tag{3-17}$$

where R is the universal gas constant, γ is adiabatic compression index.

The dissipation tensor, ε_{ij}, is modeled as

$$\varepsilon_{ij} = \frac{2}{3}\delta_{ij}\left(\rho\varepsilon + Y_M\right) \tag{3-18}$$

where Y_M is an additional dilatation dissipation term

$$Y_M = 2\rho\varepsilon M_t^2 \tag{3-19}$$

The scalar dissipation rate, ε, is computed with a model transport equation similar to that used in the standard k-ε model

$$\frac{\partial}{\partial t}(\rho\varepsilon)+\frac{\partial}{\partial x_i}(\rho\varepsilon u_i) = \frac{\partial}{\partial x_i}\left[\left(\mu+\frac{\mu_t}{\sigma_\varepsilon}\right)\frac{\partial\varepsilon}{\partial x_j}\right]C_{\varepsilon 1}\frac{1}{2}[P_{ii}+C_{\varepsilon 3}G_{ii}]\frac{\varepsilon}{k} - C_{\varepsilon 2}\rho\frac{\varepsilon^2}{k} \tag{3-20}$$

where $\sigma_\varepsilon = 1.0$, $C_{\varepsilon 1} = 1.44$, $C_{\varepsilon 2} = 1.92$, $C_{\varepsilon 3}$ is evaluated as a function of the local flow direction relative to the gravitational vector.

The production terms due to buoyancy are modeled as

$$G_{ij} = -\frac{\mu_t}{\rho \Pr_t}\left(g_i \frac{\partial \rho}{\partial x_j} + g_j \frac{\partial \rho}{\partial x_i}\right) \tag{3-21}$$

where Pr_t is the turbulent Prandtl number for energy and is 0.85.

3.1.2.2. Standard k-ε Model

Under the assumption that the flow is fully turbulent, and the effects of molecular viscosity are negligible, the turbulence kinetic energy, k, and its rate of dissipation, ε, are obtained from the following transport equations

$$\frac{\partial}{\partial t}(\rho k) + \frac{\partial}{\partial x_i}(\rho k u_i) = \frac{\partial}{\partial x_j}\left(\left(\mu + \frac{\mu_t}{\sigma_k}\right)\frac{\partial k}{\partial x_j}\right) + G_k + G_b - \rho\varepsilon - Y_M + S_k \tag{3-22}$$

$$\frac{\partial}{\partial t}(\rho\varepsilon) + \frac{\partial}{\partial x_i}(\rho\varepsilon u_i) = \frac{\partial}{\partial x_j}\left(\left(\mu + \frac{\mu_t}{\sigma_\varepsilon}\right)\frac{\partial \varepsilon}{\partial x_j}\right) + C_{\varepsilon 1}\frac{\varepsilon}{k}(G_k + G_{\varepsilon 3}G_b) - C_{\varepsilon 2}\rho\frac{\varepsilon^2}{k} + S_\varepsilon \tag{3-23}$$

In these equations, G_k represents the generation of turbulence kinetic energy due to the mean velocity gradients. G_b is the generation of turbulence kinetic energy due to buoyancy. Y_k represents the contribution of the fluctuating dilatation in compressible turbulence to the overall dissipation rate, constants $C_{\varepsilon 1}$, $C_{\varepsilon 2}$, and $C_{\varepsilon 2}$ have the same values as those in equation (3-20). σ_k and σ_ω are the turbulent Prandtl numbers for k and ε, respectively. S_k and S_ω are source terms. The turbulent (or eddy) viscosity, μ_t, is computed by combining k and ε by equation (3-14). The model constants σ_k and σ_ω have the following default values

$$\sigma_k = 1.0, \sigma_\varepsilon = 1.3 \tag{3-24}$$

The turbulent (or eddy) viscosity, μ_t, is computed by combining k and ε as given in equation (3-14)

3.1.2.3. RNG k-ε Model

The transport equations of RNG k-ε model has a similar form to the standard k-ε model

$$\frac{\partial}{\partial t}(\rho k) + \frac{\partial}{\partial x_i}(\rho k u_i) = \frac{\partial}{\partial x_j}\left(\left(\mu + \frac{\mu_t}{\sigma_k}\right)\frac{\partial k}{\partial x_j}\right) + G_k + G_b - \rho\varepsilon - Y_M + S_k \tag{3-25}$$

$$\frac{\partial}{\partial t}(\rho\varepsilon) + \frac{\partial}{\partial x_i}(\rho\varepsilon u_i) = \frac{\partial}{\partial x_j}\left(\left(\mu + \frac{\mu_t}{\sigma_\varepsilon}\right)\frac{\partial \varepsilon}{\partial x_j}\right) + C_{\varepsilon 1}\frac{\varepsilon}{k}(G_k + G_{\varepsilon 3}G_b) - C_{\varepsilon 2}\rho\frac{\varepsilon^2}{k} - R_\varepsilon + S_\varepsilon \tag{3-26}$$

The turbulent (or eddy) viscosity, μ_t, is computed by combining k and ε as given in equation (3-14), but the constant C_μ= 0.0845.

The production of turbulent kinetic energy, G_k, is modeled as

$$G_k = \mu_t S^2 \tag{3-27}$$

where S is the modulus of the mean rate-of-strain tensor, defined as

$$S \equiv \sqrt{2S_{ij}S_{ij}} \tag{3-28}$$

and

$$S_{ij} = \frac{1}{2}\left(\frac{\partial u_j}{\partial x_i} + \frac{\partial u_i}{\partial x_j}\right) \tag{3-29}$$

The generation of turbulence due to buoyancy is given by

$$G_b = -g_i \frac{\mu_t}{\rho \Pr_t} \frac{\partial \rho}{\partial x_i} \tag{3-30}$$

where Pr_t is the turbulent Prandtl number for energy and g_i is the component of the gravitational vector in the i_{th} direction. For the standard k-ε model, the default value of Pr_t is 0.85. In the case of the RNG k-ε model, $Pr_t = 1/\alpha$, where α is given by

$$\left|\frac{\alpha - 1.3929}{\alpha_0 - 1.3929}\right|^{0.6321} \left|\frac{\alpha + 2.3929}{\alpha_0 + 2.3929}\right|^{0.3679} = \frac{\mu}{\mu_{\mathrm{eff}}} \tag{3-31}$$

where

$$\alpha_0 = \frac{1}{\Pr} = \frac{k}{\mu c_P} \tag{3-32}$$

The degree to which ε is affected by the buoyancy is determined by the constant $C_{\mu 3}$ and is calculated by

$$C_{\mu 3} = \tanh\left|\frac{v}{u}\right| \tag{3-33}$$

where v is the component of the flow velocity parallel to the gravitational vector and u is the component of the flow velocity perpendicular to the gravitational vector. In this way, $C_{\mu 3}$ will become 1 for buoyant shear layers for which the main flow direction is aligned with the direction of gravity. For buoyant shear layers that are perpendicular to the gravitational vector, $C_{\mu 3}$ will become zero.

The main difference between the RNG and standard k-ε models lies in the additional term in the ε equation given by

$$R_\varepsilon = \frac{C_\mu \rho \eta^3 (1 - \eta/\eta_0)}{1 + \beta \eta^3} \frac{\varepsilon^2}{k} \tag{3-34}$$

where $\eta = Sk/\varepsilon$, $\eta_0 = 4.38$, $\beta = 0.012$.

3.1.2.4. Realizable k-ε *Model*

The transport equations for k and ε in the realizable k-ε model are

$$\frac{\partial}{\partial t}(\rho k) + \frac{\partial}{\partial x_i}(\rho k u_i) = \frac{\partial}{\partial x_j}\left(\left(\mu + \frac{\mu_t}{\sigma_k}\right)\frac{\partial k}{\partial x_j}\right) + G_k + G_b - \rho\varepsilon - Y_M + S_k \tag{3-35}$$

$$\frac{\partial}{\partial t}(\rho \varepsilon) + \frac{\partial}{\partial x_i}(\rho \varepsilon u_i) = \frac{\partial}{\partial x_j}\left(\left(\mu + \frac{\mu_t}{\sigma_\varepsilon}\right)\frac{\partial \varepsilon}{\partial x_j}\right) + \rho C_1 S\varepsilon - \rho C_2 \frac{\varepsilon^2}{k + \sqrt{\nu\varepsilon}} + C_{\varepsilon 1}\frac{\varepsilon}{k}C_{\varepsilon 3}G_b + S_\varepsilon \tag{3-36}$$

where

$$C_1 = \max\left[0.43, \frac{\eta}{\eta + 5}\right] \tag{3-37}$$

$$\eta = S\frac{k}{\varepsilon} \tag{3-38}$$

The eddy viscosity is computed from

$$\mu_t = \rho C_\mu \frac{k^2}{\varepsilon} \tag{3-39}$$

and

$$C_\mu = \frac{1}{A_0 + A_s \dfrac{kU^*}{\varepsilon}} \tag{3-40}$$

$$U^* = \sqrt{S_{ij}S_{ij} + \overline{\Omega}_{ij}\overline{\Omega}_{ij}} \tag{3-41}$$

$$\overline{\Omega}_{ij} = \Omega_{ij} - 2\varepsilon_{ij}k\omega_{ij} \quad (3\text{-}42)$$

$$\Omega_{ij} = \widehat{\Omega}_{ij} - \varepsilon_{ij}k\omega_{k} \quad (3\text{-}43)$$

where $\widehat{\Omega}_{ij}$ is the mean rate-of-rotation tensor viewed in a rotating reference frame with the angular velocity ω_k. The model constants A_0 =4.04 and A_s is given by

$$A_s = \sqrt{6}\cos\phi \quad (3\text{-}44)$$

where

$$\phi = \frac{1}{3}\cos^{-1}\left(\sqrt{6}W\right) \quad (3\text{-}45)$$

$$W = \frac{S_{ij}S_{jk}S_{ki}}{\overline{S}^3} \quad (3\text{-}46)$$

$$\overline{S} = \sqrt{S_{ij}S_{ij}} \quad (3\text{-}47)$$

The model constants $C_{\varepsilon 1}$, C_2, σ_k, and σ_ε are

$$C_{\varepsilon 1} = 1.44, C_2 = 1.9, \sigma_k = 1.0, \sigma_\varepsilon = 1.2 \quad (3\text{-}48)$$

3.1.2.5. Standard k-ω Model

The standard k-ω model is an empirical model based on model transport equations for the turbulence kinetic energy k and the specific dissipation rate ω, which can also be thought of as the ratio of ε to k, for low-Reynolds-number effects.

$$\frac{\partial}{\partial t}(\rho k) + \frac{\partial}{\partial x_i}(\rho k u_i) = \frac{\partial}{\partial x_j}\left(\Gamma_k \frac{\partial k}{\partial x_j}\right) + G_k - Y_k + S_k \quad (3\text{-}49)$$

$$\frac{\partial}{\partial t}(\rho \omega) + \frac{\partial}{\partial x_i}(\rho \omega u_i) = \frac{\partial}{\partial x_j}\left(\Gamma_\omega \frac{\partial \omega}{\partial x_j}\right) + G_\omega - Y_\omega + S_\omega \quad (3\text{-}50)$$

where Γ_k and Γ_ω represent the effective diffusivity of k and ε, respectively, Y_k and Y_ω represent the dissipation of k and ε due to turbulence.

The effective diffusivities for the k-ω model are given by

$$\Gamma_k = \mu + \frac{\mu_t}{\sigma_k} \quad (3\text{-}51)$$

$$\Gamma_w = \mu + \frac{\mu_t}{\sigma_w} \tag{3-52}$$

where σ_k and σ_ω are the turbulent Prandtl numbers for k and ω, respectively. The turbulent viscosity, μ_t, is computed by combining k and ω as follows

$$\mu_t = \alpha^* \frac{\rho \kappa}{w} \tag{3-53}$$

The coefficient α^* damps the turbulent viscosity causing a low-Reynolds-number correction and is given by

$$\alpha^* = \alpha^*_\infty \left(\frac{\alpha^*_0 + \frac{\mathrm{Re}_t}{6}}{1 + \frac{\mathrm{Re}_t}{6}} \right) \tag{3-54}$$

where

$$\mathrm{Re}_t = \frac{\rho \kappa}{\mu w} \tag{3-55}$$

$$\alpha^*_0 = 0.024 \tag{3-56}$$

in the form of the k-ω model at high-Reynolds-number, $\alpha^* = \alpha^*_\infty = 1$.

The production of ω is given by

$$G_w = \alpha \frac{w}{k} G_k \tag{3-57}$$

where

$$\alpha = \frac{\alpha_\infty}{\alpha^*} \left(\frac{\alpha_0 + \mathrm{Re}_t / \mathrm{R}_\omega}{1 + \mathrm{Re}_t / \mathrm{R}_\omega} \right) \tag{3-58}$$

where $R_\omega = 2.95$.

The dissipation of k is given by

$$\mathrm{Y}_\mathrm{k} = \rho \beta^* \mathrm{f}_{\beta *} \mathrm{k} \omega \tag{3-59}$$

where

$$f_{\beta^*} = \begin{cases} 1 & \chi_k \le 0 \\ \dfrac{\frac{1}{1+680\chi_k^2}}{1+400\chi_k^2} & \chi_k > 0 \end{cases} \tag{3-60}$$

where

$$\chi_k = \frac{1}{\omega^3}\frac{\partial k}{\partial x_j}\frac{\partial \omega}{\partial x_j} \tag{3-61}$$

and

$$\beta^* = \beta^*_i\left(1+1.5F(M_t)\right) \tag{3-62}$$

$$\beta^*_i = 0.09\left(\frac{4/15+(Re_t/8)^4}{1+(Re_t/8)^4}\right) \tag{3-63}$$

The dissipation of ω is given by

$$Y_\omega = \rho\beta f_\beta \omega^2 \tag{3-64}$$

where

$$f_\beta = \frac{1+70\chi_\omega}{1+80\chi_\omega} \tag{3-65}$$

$$\chi_\omega = \left|\frac{\Omega_{ij}\Omega_{jk}S_{ki}}{(\beta^*_\infty\omega)^3}\right| \tag{3-66}$$

$$\Omega_{ij} = \frac{1}{2}\left(\frac{\partial u_i}{\partial x_j} - \frac{\partial u_j}{\partial x_i}\right) \tag{3-67}$$

$$\beta = 0.072\left|1-1.5\frac{\beta^*_i}{0.072}F(M_t)\right| \tag{3-68}$$

The compressibility function, $F(M_t)$, is given by

$$F(M_t) = \begin{cases} 0 & M_t \le M_{t0} \\ M_t^2 - M_{t0}^2 & M_t > M_{t0} \end{cases} \tag{3-69}$$

where

$$M_t^2 = \frac{2k}{a^2} \tag{3-70}$$

and M_{t0}=0.25, $a = \sqrt{\gamma RT}$.

Model constants are

$$\alpha^*_\infty = 1, \alpha_\infty = 0.52, \alpha_0 = 1/9, \beta^*_\infty = 0.09, \beta_i = 0.072, R_\omega = 2.95, M_{t0} = 0.25$$
$$\sigma_k = 2.0, \sigma_\omega = 2.0 \tag{3-71}$$

3.1.3. Radiation Model

The radiative heat flux vector in a gray medium can be given by

$$q_r = -\Gamma \nabla G \tag{3-72}$$

where

$$\Gamma = \frac{1}{\left(3\left(a+\sigma_s\right)-C\sigma_s\right)} \tag{3-73}$$

The transport equation for G is

$$\nabla\left(\Gamma \nabla G\right) - aG + 4a\sigma T^4 = S_G \tag{3-74}$$

where σ is the Stefan-Boltzmann constant and S_G is radiation source term.

Combining equations (3-72) and (3-74), the following equation is obtained

$$-\nabla q_r = aG - 4a\sigma T^4 \tag{3-75}$$

The expression for $-\nabla q_r$ can be directly substituted into the energy equation to account for heat sources (or sinks) due to radiation.

3.1.4. Chemical Reaction Model

Most fuels are fast burning, and the overall rate of reaction is controlled by turbulent mixing. In non-premixed flames, turbulence slowly convects/mixes fuel and oxidizer into the reaction zones where they burn quickly. In such case, the combustion is said to be mixing-limited, chemical kinetic rates can be safely neglected.

In the eddy-dissipation model, a turbulence-chemistry interaction model, the net rate of production of species i due to reaction r, $R_{i,r}$, is given by the smaller (i.e., limiting value) of the two expressions below

$$R_{i,r} = v_{i,r}^{'} M_{w,i} A \rho \frac{\varepsilon}{\kappa} \min_{\Re} \left(\frac{Y_{\Re}}{v_{\Re,r}^{'} M_{w,\Re}} \right) \tag{3-76}$$

$$R_{i,r} = v_{i,r}^{'} M_{w,i} AB \rho \frac{\varepsilon}{\kappa} \frac{\sum_{P} Y_{P}}{\sum_{j}^{N} v_{j,r}^{''} M_{w,j}} \tag{3-77}$$

where Y_P is the mass fraction of any product species, P. $Y_{\mathcal{R}}$ is the mass fraction of a particular reactant, $\mathcal{R}$, A is an empirical constant equal to 4.0 and B is an empirical constant equal to 0.5.

In equations (3-76) and (3-77), the chemical reaction rate is governed by the large eddy mixing time scale, k/ε, as in the eddy-breakup model. Combustion proceeds whenever turbulence is present ($k/\varepsilon > 0$), and an ignition source is not required to initiate combustion. This is usually acceptable for non-premixed flames.

3.2. Simulation of Diffusion Flames

3.2.1. Computational Grid

Since the combustion chamber, which is discussed in Chapter 1 for laminar diffusion flames, is symmetrical around the central axis, a two-dimensional computational grid with a radial symmetry axis is constructed for simulation, as shown in Figure 3-1. The grid has 283 and 87 nodes in the axial and radial directions, respectively, and is refined by gradient adaptation approach in the area above the burner where gradients are expected to be higher.

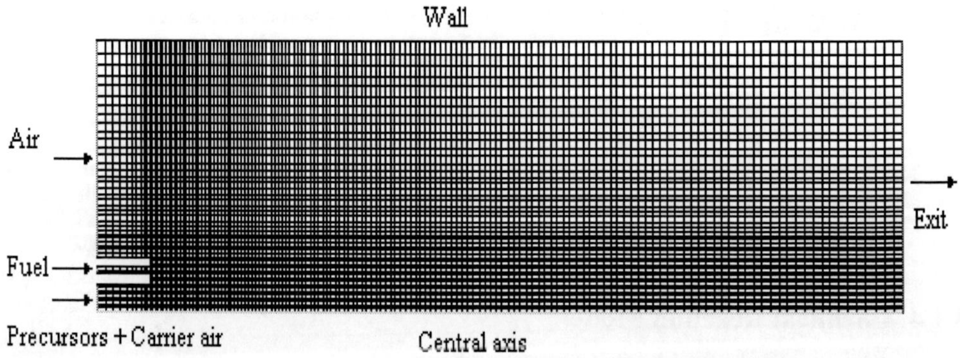

Figure 3.1. Illustration of computational grid.

3.2.2. Physical Properties

Materials in the combustor are C_3H_8, O_2, CO_2, H_2O and N_2. To minimize numerical error, N_2 is selected as the last specie as N_2 is the overall largest mass fraction. The overall chemical reaction is taken as

$$C_3H_8 + O_2 \rightarrow CO_2 + H_2O \tag{3-78}$$

The properties include the following

- density and/or molecular weights
- viscosity
- heat capacity
- thermal conductivity
- diffusion coefficients
- standard state enthalpies
- adsorption coefficient for irradiation, which is taken as 12.731/m
- scattering coefficient for irradiation, which is taken as 0.

The properties may be temperature-dependent and/or composition-dependent, with temperature dependence based on a polynomial, piecewise-linear, or piecewise-polynomial function and individual component properties. Parameters for the properties are given in Table 3-1.

3.2.3. Boundary Conditions

The turbulence length scale, l, is a physical quantity related to the size of the large eddies that contain the energy in turbulent flows. In fully-developed duct flows, l is restricted by the size of the duct, since the turbulent eddies cannot be larger than the duct. An approximate relationship between l and the physical size of the duct is

$$\ell = 0.07 L \tag{3-79}$$

where L is the relevant dimension of the duct.The factor of 0.07 is based on the maximum value of the mixing length in fully-developed turbulent pipe flow, where L is the diameter of the pipe. In a channel of non-circular cross-section, L is the hydraulic diameter.

The scalar dissipation rate, ε, is determined by

$$\varepsilon = C_\mu^{3/4}\left(\frac{k^{3/2}}{L}\right) \tag{3-80}$$

where C_μ is an empirical constant specified in the turbulence model (approximately 0.09).

Table 3.1. Parameters of physical properties

Gas properties	
heat capacity	J/kg·K
CO_2	429.900+1.874T-1.970E-3T^2+1.297E-6T^3-4.001E-10T^4(300～1000K) 841.38+0.5932T-2.400E-4T^2+4.5227E-8T^3-3.150E-12T^4(1000～5000K)
N_2	938.899+0.302T-8.109E-5T^2+8.264E-9T^3-1.537E-13T^4
O_2	811.180+0.411T-1.750E-4T^2+3.758E-8T^4-2.974E-12T^4
C_3H_8	169.111+5.032T+1.024E-3T^2-4.008E-6T^3+1.743E-9T^4(300～1000K) 1418.84T+3.562T-1.185E-3T^2+1.731E-7T^3-9.074E-12T^4(1000～5000K)
H_2O	1609.791+0.7405T-9.130E-6T^2-3.814E-8T^3+4.802E-12T^4
thermal conductivity	W/m·K
CO_2	-0.0075+7E-5T-3E-8T^2-5E-11T^3-1E-14T^4
N_2	0.00474+7.272E-5T-1.122E-8T^2+1.455E-12T^3-7.872E-17T^4
O_2	0.00392+8.0812E-5T-1.354E-8T^2+2.220E-12T^3-1.416E-16T^4
C_3H_8	-0.0019+3E-5T+1E-7T^2+1E-11T^3-4E-15T^4
H_2O	-0.00791+6.8813E-5T+4.49E-8T^2-9.0999E-12T^3+6.173E-16T^4
viscosity	Pa·s
CO_2	-1.549E-6+6.21E-8T-2E-11T^2+5E-15T^3+3E-19T^4
N_2	7.473E-6+4.084E-8T-8.245E-12T^2+1.306E-15T^3-8.178E-17T^4
O_2	7.879E-6+4.925E-8T-9.852E-12T^2+1.527E-15T^3-9.426E-20T^4
C_3H_8	-1.601E-7+2.79E-8T+4E-12T^2-1E-14T^3+6E-18T^4
H_2O	-4.419E-6+4.6876E-8T-5.389E-12T^2+3.2029E-16T^3+4.919E-22T^4
diffusivity	m^2/s
C_3H_8 O_2	-1.37E-6+2E-08T+9E-11T^2-1E-14T^3+2E-18T^4
C_3H_8 CO_2	-1.1E-6+1E-08T+7E-11T^2-1E-14T^3+1E-18T^4
C_3H_8 H_2O	-1.76E-6+2E-08T+1E-10T^2-2E-14T^3+2E-18T^4
C_3H_8 N_2	-1E-6+2E-08T+9E-11T^2-1E-14T^3+2E-18T^4
O_2 CO_2	-2E-6+2E-08T+1E-10T^2-2E-14T^3+3E-18T^4
O_2 H_2O	-3.14E-6+4E-08T+2E-10T^2-3E-14T^3+4E-18T^4
O_2 N_2	-2E-6+3E-08T+2E-10T^2-2E-14T^3+4E-18T^4
CO_2 H_2O	-3E-6+3E-08T+2E-10T^2-2E-14T^3+3E-18T^4
CO_2 N_2	-2E-6+2E-08T+1E-10T^2-2E-14T^3+3E-18T^4
H_2O N_2	-3.14E-6+4E-08T+2E-10T^2-3E-14T^3+4E-18T^4

The turbulent kinetic energy, k, is determined by

$$k = (0.5 - 1.5\%)\left(\frac{1}{2}u_{avg}^2\right) \tag{3-81}$$

where u_{avg} is the mean flow velocity.

The specific dissipation rate, ω, is determined by

$$\omega = \frac{\sqrt{k}}{C_\mu^{1/4} l} \tag{3-82}$$

The wall temperature is assumed to be unchanged with height and taken as 500K in the calculations, which is close to the measured value in the experiments. The pressure and gas velocity at the inlet are taken as constant according to the experimental values. The gauge pressure at the outlet is taken as zero.

In the flame model, the combustion rates are controlled by the Eddy-Dissipation model, and the turbulence is simulated by Standard and RNG k-ε, and Standard k-ω models and controlled by equilibrium and non-equilibrium wall functions in the wall region. SIMPLE method is selected for pressure-velocity coupling and the convection term is discretized with fist-order upwind scheme [1]. Radiant energy absorption and irradiant are incorporated in the enthalpy balance by P1 model [1].

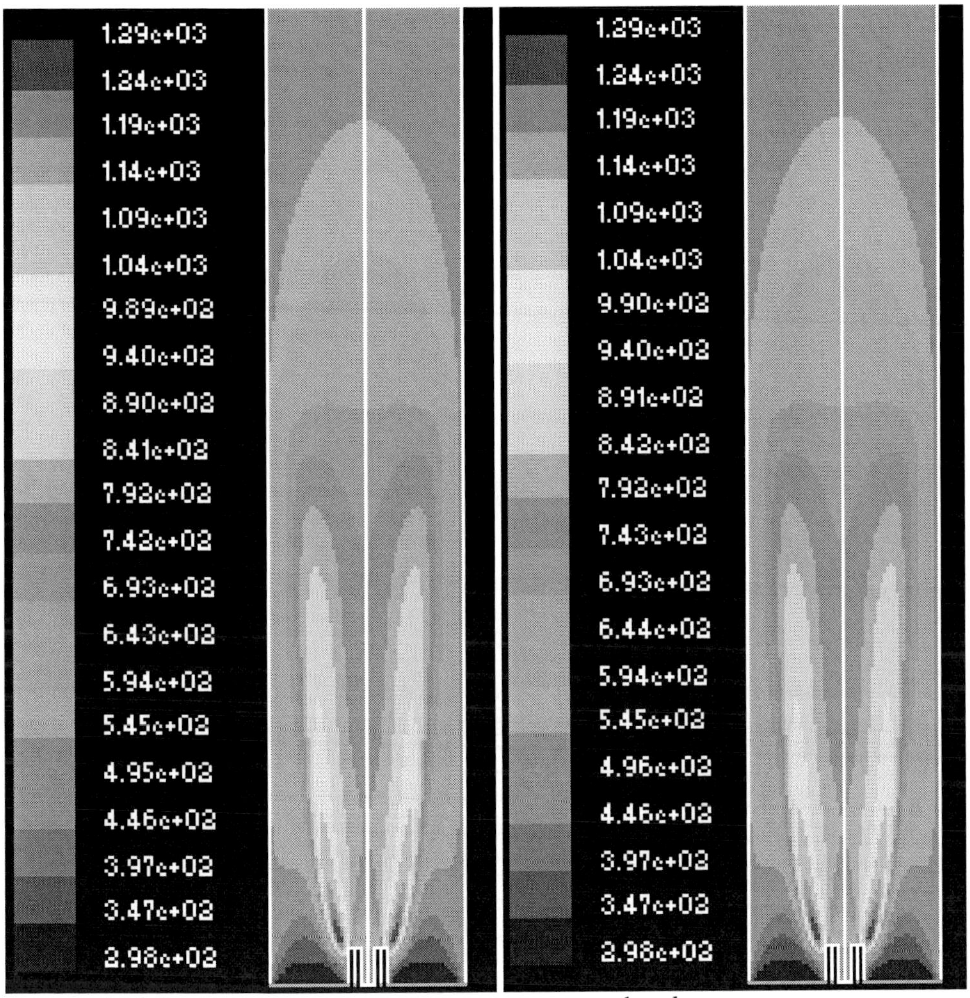

high *k* low *k*

Figure 3.2. Comparison of simulated temperature with different *k* values near wall region.

3.2.4. Comparison between Different Turbulent Models

3.2.4.1. Effects of Turbulent Kinetic Energy, k

The turbulent kinetic energy, *k*, near wall region is determined by equation (3-81). When the coefficient in equation (3-81) is taken as 0.5% and 1.5%, respectively, the effects of turbulent kinetic energy, *k*, on simulation results of laminar diffusion flame is examined. The operational conditions and measurement results of the flame simulated are listed in Table 3-2. The simulated temperature contour plots by Standard k-ε controlled by non-equilibrium wall functions in the wall region for the two different k values on the boundary are shown in Figure 3-2. One could see that the effects of the k values near wall region on flame temperature are minor.

Table 3.2. Operational conditions and measurement results of laminar diffusion flame

Pressure	Propane flow rate	Air flow rate	Flame length	Tempe-rature	inlet temperature
1.05atm	0.000777mol/s	0.01533mol/s	0.335m	1150K	298K

From the table, propane and air speeds in the combustor are

$$v_{c_3h_8} = \frac{0.000777 \times 24.45 \times 10^{-3}}{\pi(r_2 - r_1)} \tag{3-83}$$

$$v_{o_2} = \frac{0.015133 \times 24.45 \times 10^{-3}}{\pi(R - r_2)} \tag{3-84}$$

where r_1=6.2mm, r_2=6.9mm, and R=51.5mm.

3.2.4.2. Effects of Wall Treatments

The simulated temperature contour plots by Standard k-ε controlled by equilibrium and non-equilibrium wall functions in the wall region are shown in Figure 3-3. Equilibrium wall functions are for the flow along the wall. Non-equilibrium wall functions are suitable for more complicated flows such as flow separation, return flow, impact flow and etc.. One could see that the simulated flame temperature with non-equilibrium wall functions is more close to the measurement results, slight lower than that simulated with equilibrium wall functions.

3.2.4.3. Effects of Different Turbulent Models

A comparison on temperature contour plots simulated by Standard, RNG and Realizable k-ε models controlled by non-equilibrium wall functions is shown in Figure 3-4. It shows that Standard and RNG k-ε models give similar predictions on flame configuration and temperature distributions, however, Realizable k-ε model gives a longer flame. Measured temperature is about 1150K as given in Table 3-2, the maximum temperature simulated by Standard, RNG and Realizable k-ε models are 1290K, 1360K, 1340K, respectively.

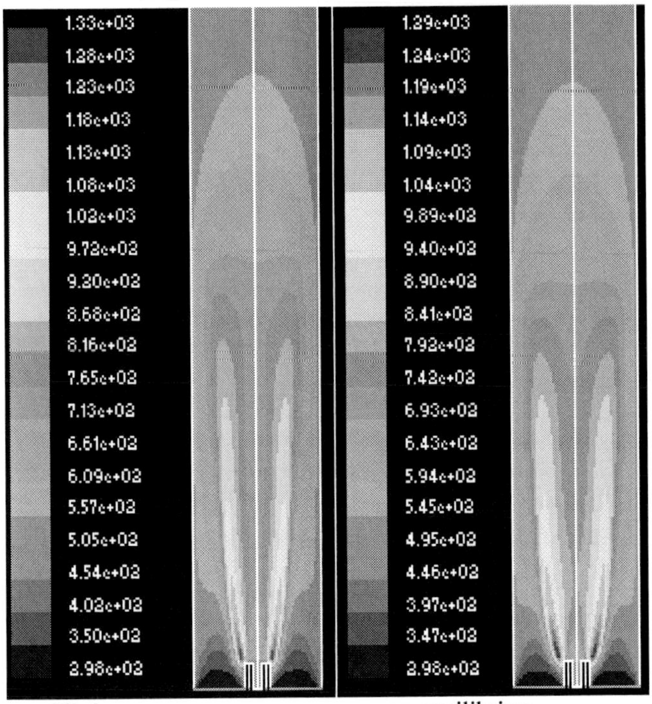

equilibrium non-equilibrium

Figure 3.3. Comparison of simulated temperature with different wall functions.

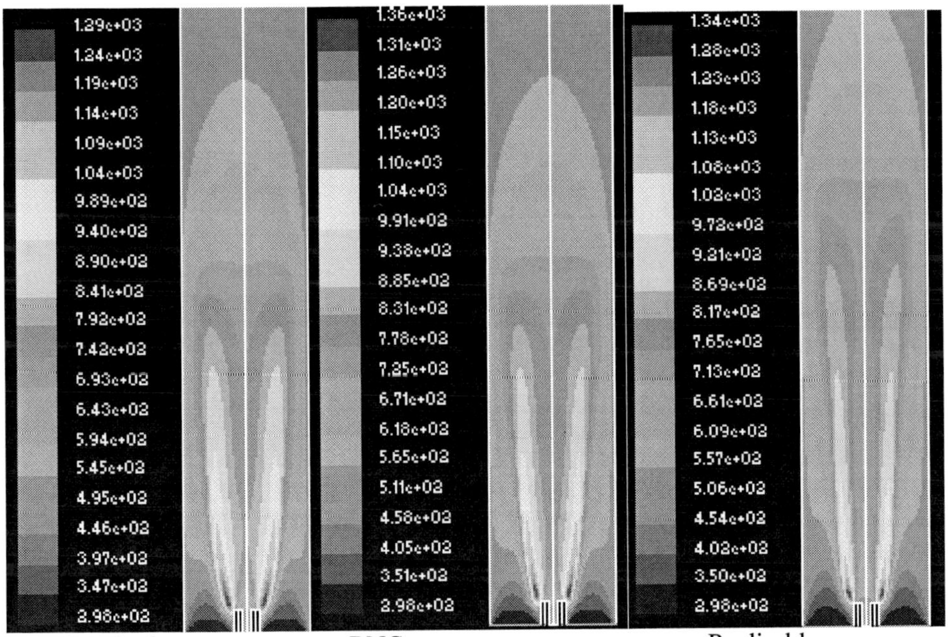

Standard RNG Realizable

Figure 3.4. Comparison between three different k-ε models.

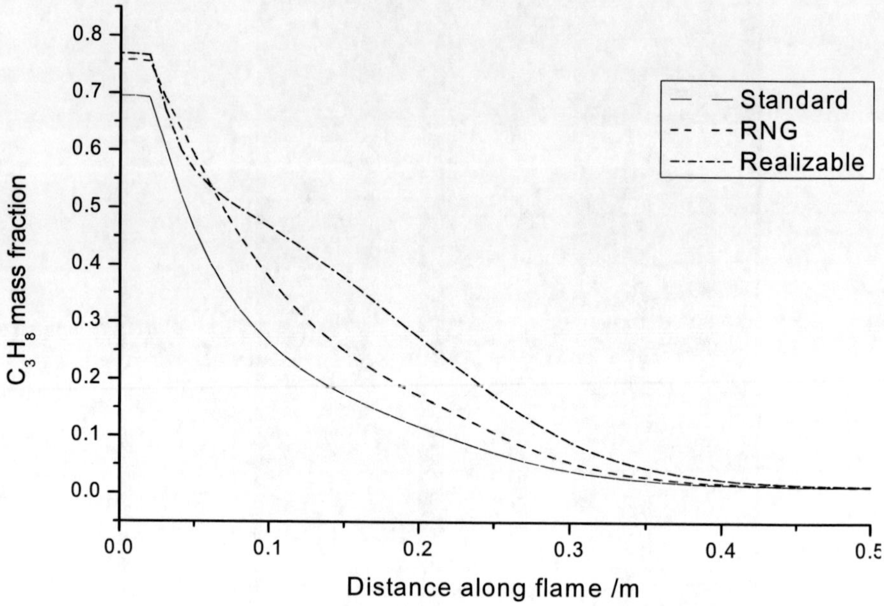

Figure 3.5. C_3H_8 mass fractions on the central axis simulated by Standard, RNG and Realizable k-ε models.

A comparison on C3H8 mass fraction on the central axis simulated by Standard, RNG and Realizable k-ε models controlled by non-equilibrium wall functions is shown in Figure 3-5. It again shows that Realizable k-ε model gives a longer flame. The flame length obtained from Figure 3-5 by the distance when C_3H_8 mass fraction along central axis is zero is about 0.35m for Standard and RNG k-ε models, which is close to the measured flame length of 0.335m as given in Table 3-2, and is about 0.4m for Realizable k-ε model.

3.2.5. Simulation by Detailed Chemical Reactions

3.2.5.1. Universal Physics Equations

The net rate of production of species i by chemical reaction, R_i, in species transport equation (3-10) contains the sum of all species participating chemical reactions with species i

$$R_i = M_{w,i} \sum_{i=1}^{N_r} \hat{R}_{i,r} \tag{3-85}$$

where $M_{w,i}$ is the molecular weight of species i, and $\hat{R}_{i,r}$ is the production/destruction rate of species i in reaction r

$$\hat{R}_{i,r} = A_r T^{\beta_r} e^{\frac{-E_r}{RT}} \tag{3-86}$$

where A_r is the pre-exponential factor, β_r is temperature index, E_r is the activation energy (J/kmol), and R is universal gas constant (J/kmol/K).

The energy transportation equation is

$$\frac{\partial}{\partial t}(\rho E)+\nabla\bullet(\vec{v}(\rho E+p))=\nabla\bullet(k_{eff}\nabla T-\sum h_j\vec{J}_j+(\vec{\tau}_{eff}\bullet\vec{v}))+S_j \tag{3-87}$$

The source term S_j in energy transportation equation is

$$S_j=-j\sum_j\left(\frac{h_j^0}{M_j}+\int_{T_{ref}}^{T}c_{pj}dT\right)R_j \tag{3-88}$$

where h_j^0 is the standard enthalpy of formation of species j, R_j is the volumetric heat release rate of species j.

Then the universal physics equations such as mass, momentum, energy can be written as

$$\frac{\partial}{\partial x}(\rho u\varphi)+\frac{\partial}{\partial x}(\rho v\varphi)=\frac{\partial}{\partial y}\left(\Gamma_\varphi\frac{\partial\varphi}{\partial x}\right)+\frac{\partial}{\partial y}\left(\Gamma_\varphi\frac{\partial\varphi}{\partial y}\right)+S_\varphi \tag{3-89}$$

Table 3.3. φ, Γ_φ, S_φ in models of calculation

equation	φ	Γ_φ	S_φ
Mass conservation	1	0	0
Momentum conservation in x direction	u	μ_e	$-\frac{\partial p}{\partial x}+\frac{\partial}{\partial x}(\mu_e\frac{\partial u}{\partial x})+\frac{\partial}{\partial y}(\mu_e\frac{\partial v}{\partial x})$
Momentum conservation in y direction	v	μ_e	$-\frac{\partial p}{\partial x}+\frac{\partial}{\partial y}(\mu_e\frac{\partial u}{\partial y})+\frac{\partial}{\partial x}(\mu_e\frac{\partial v}{\partial y})$
Turbulent kinetic energy	k	$\frac{\mu_e}{\sigma_k}$	$G_k-\rho t$
Scalar dissipation rate	ε	$\frac{\mu_e}{\varepsilon}$	$\frac{\varepsilon}{k}(c_1G_K-c_2\rho\varepsilon)$
Species	Y_s	$\frac{\mu_e}{\sigma_Y}$	R_i
Energy	H	$\frac{\mu_e}{\sigma_k}$	$-R_iQ$

Table 3.4. Propane combustion rate constants in cm^3 moles kcal units

$$\Re = A_r T^{\beta_r} \exp(-E_r / RT)$$

	Reaction	$\log A_r$	β_r	E_r
1	$C_3H_8 \rightarrow C_2H_5+CH_3$	16.6	0	84.8
2	$C_3H_8+O_2 \rightarrow C_3H_7+HO_2$	13.0	0	52.0
3	$C_3H_8+O_2 \rightarrow H_7C_3+HO_2$	12.5	0	52.0
4	$CH_3+C_3H_8 \rightarrow C_3H_7+CH_4$	12.5	0	11.4
5	$CH_3+C_3H_8 \rightarrow C_3H_7+CH_4$	11.9	0	8.6
6	$H+C_3H_8 \rightarrow C_3H_7+H_2$	13.7	0	9.1
7	$H+C_3H_8 \rightarrow H_7C_3+H_2$	13.1	0	6.4
8	$O+C_3H_8 \rightarrow C_3H_7+OH$	13.6	0	11.1
9	$O+C_3H_8 \rightarrow H_7C_3+OH$	13.0	0	8.4
10	$OH+C_3H_8 \rightarrow C_3H_7+H_2O$	4.7	2.7	0.1
11	$OH+C_3H_8 \rightarrow H_7C_3+H_2O$	3.7	2.8	-1.7
12	$HO_2+C_3H_8 \rightarrow C_3H_7+H_2O_2$	11.8	0	10.5
13	$HO_2+C_3H_8 \rightarrow H_7C_3+H_2O_2$	11.3	0	10.5
14	$C_3H_7 \rightarrow C_2H_4+CH_3$	13.6	0	33.2
15	$H_7C_3 \rightarrow C_2H_4+CH_3$	12.0	0	34.5
16	$C_3H_7 \rightarrow C_3H_6+H$	13.8	0	38.0
17	$H_7C_3 \rightarrow C_3H_6+H$	14.3	0	41.3
18	$C_3H_7+O_2 \rightarrow C_3H_6+HO_2$	12.0	0	5.0
19	$H_7C_3+O_2 \rightarrow C_3H_6+HO_2$	12.0	0	5.0
20	$CH_3+C_3H_6 \rightarrow C_3H_5+CH_4$	12.4	0	5.4
21	$H+C_3H_6 \rightarrow C_2H_4+CH_3$	12.6	0	0
22	$H+C_3H_6 \rightarrow C_3H_5+H_2$	13.7	0	5.0
23	$OH+C_3H_5 \rightarrow C_2H_4+CH_2O$	13.0	0	0
24	$O+C_3H_6 \rightarrow C_2H_5+HCO$	13.4	0	2.7
25	$OH+C_3H_6 \rightarrow C_2H_5+CH_2O$	13.0	0	0
26	$C_2H_5 \rightarrow C_2H_4+H$	13.5	0	40.7
27	$M+CH_4 \rightarrow CH_3+H+M$	17.1	0	88.4
28	$H+CH_4 \rightarrow CH_3+H_2$	14.8	0	15.1
29	$O+CH_4 \rightarrow CH_3+OH$	14.6	0	14.0
30	$OH+CH_4 \rightarrow CH_3+H_2O$	13.5	0	6.0
31	$CH_3+O \rightarrow CH_2O+H$	14.1	0	2.0
32	$CH_3+O_2 \rightarrow CH_3O+O$	13.4	0	29.0
33	$CH_3+HO_2 \rightarrow CH_3O+OH$	13.2	0	0
34	$CH_3+OH \rightarrow CH_2O+H_2$	12.6	0	0
35	$M+CH_3O \rightarrow CH_2O+H$	13.7	0	21.0
36	$CH_3O+O_2 \rightarrow CH_2O+HO_2$	12.0	0	6.0
37	$CH_3+CH_3 \rightarrow C_2H_6$	12.8	0	-1.0
38	$CH_3+C_2H_6 \rightarrow C_2H_5+CH_4$	-0.3	4.0	8.3
39	H+C2H6→C2H5+H2	7.5	2.0	6.9

	Reaction	logAr	□r	Er
40	$O+C_2H_6 \rightarrow C_2H_5+OH$	13.2	0	6.1
41	$OH+C_2H_6 \rightarrow C_2H_5+H_2O$	13.8	0	3.6
42	$HO_2+C_2H_6 \rightarrow C_2H_5+H_2O_2$	12.0	0	14.0
43	$C_2H_5+O_2 \rightarrow C_2H_4+HO_2$	12.0	0	5.0
44	$H+C_2H_5 \rightarrow 2CH_3$	13.0	0	0
45	$H+C_2H_5 \rightarrow H_2+C_2H_4$	13.7	0	0
46	$H+C_2H_4 \rightarrow C_2H_3+H_2$	14.0	0	8.5
47	$O+C_2H_4 \rightarrow CH_3+HCO$	13.4	0	2.7
48	$OH+C_2H_4 \rightarrow C_2H_3+H_2O$	14.0	0	3.5
49	$M+C_2H_3 \rightarrow C_2H_2+H+M$	14.9	0	31.5
50	$H+C_2H_2 \rightarrow C_2H+H_2$	14.3	0	19.0
51	$O+C_2H_2 \rightarrow CH_2+CO$	13.7	0	3.7
52	$OH+C_2H_2 \rightarrow C_2H+H_2O$	12.8	0	7.0
53	$C_2H+O_2 \rightarrow HCO+CO$	13.0	0	7.0
54	$CH_2+O_2 \rightarrow HCO+OH$	14.0	0	3.7
55	$H+CH_2O \rightarrow HCO+H_2$	14.5	0	10.5
56	$O+CH_2O \rightarrow HCO+OH$	13.3	0	3.1
57	$OH+CH_2O \rightarrow HCO+H_2O$	12.9	0	0.2
58	$HO_2+CH2O \rightarrow HCO+H_2O_2$	12.1	0	8.0
59	$M+HCO \rightarrow CO+H+M$	14.7	0	19.0
60	$H+HCO \rightarrow CO+H_2$	14.3	0	0
61	$O+HCO \rightarrow CO+OH$	14.0	0	0
62	$OH+HCO \rightarrow CO+H_2O$	14.0	0	0
63	$HCO+O_2 \rightarrow CO+HO_2$	12.5	0	7.0
64	$OH+CO \rightarrow CO_2+H$	7.2	1.3	-0.7
65	$HO_2+CO \rightarrow CO_2+OH$	14.2	0	23.6
66	$CO+O+M \rightarrow CO_2+M$	15.4	0	4.4
67	$H+O_2 \rightarrow OH+O$	17.1	-0.9	16.6
68	$OH+H_2 \rightarrow H_2O+H$	7.5	1.8	3.0
69	$O+H_2 \rightarrow OH+H$	14.3	0	13.8
70	$OH+OH \rightarrow H_2O+O$	13.7	0	7.0
71	$H+OH+M \rightarrow H_2O+M$	22.3	-2.0	0
72	$H+H+M \rightarrow H_2+M$	17.8	-1.0	0
73	$H+O_2+M \rightarrow HO_2+M$	18.5	-1.0	0
74	$OH+HO_2 \rightarrow H_2O+O_2$	13.7	0	1.0
75	$H+HO_2 \rightarrow H_2+O_2$	13.4	0	0.7
76	$H+HO_2 \rightarrow 2OH$	14.7	0	1.8
77	$O+HO_2 \rightarrow OH+O_2$	13.7	0	1.0
78	$HO_2+HO_2 \rightarrow H_2O_2+O_2$	12.3	0	0
79	$HO_2+H_2 \rightarrow H_2O_2+H$	11.5	0	18.7
80	$OH+H_2O_2 \rightarrow HO_2+H_2O$	13.0	0	1.9
81	$M+H_2O_2 \rightarrow 2OH+M$	17.1	0	45.5
82	$H+H_2O_2 \rightarrow H_2O+OH$	12.7	0	9.9
83	$O+H_2O_2 \rightarrow HO_2+OH$	13.3	0	5.9

where φ is general dependent flux variable, Γ_φ is transportation coefficient, S_φ is source term. Standings and meaning of φ, Γ_φ, S_φ is tabulated in Table 3-3.

3.2.5.2. Propane Combustion Mechanisms

The combustion process in propane/air laminar diffusion flame is simulated from Casimir's propane combustion mechanisms [2] by defining reaction components and intermediates as self-defined scalars in CFD code Fluent. The Casimir propane combustion mechanisms contain 27 components and intermediates and involve 83 chemical reactions. The rate constants of each reaction are listed in Table 3-4.

3.2.5.3. Simulation Conditions

Materials in the combustor contains 27 components of C_3H_8,O_2, OH, H_2O_2, C_3H_6, H, C_2H_3, HO_2, C_2H_4, C_2H_2, CH_2O, C_2H_5, H_2O, C_2H, HCO, CO, CH_3, C_2H_6, CH_2, CH_4, CO_2, O, C_3H_7, H_2, CH_3O, H_7C_3, C_3H_5 and N_2. To minimize numerical error, N_2 is selected as the last specie as N_2 is the overall largest mass fraction.

The properties may be temperature-dependent and/or composition-dependent, with temperature dependence based on a polynomial, piecewise-linear, or piecewise-polynomial function and individual component properties. Diffusion coefficient between two different species A and B is calculated by

$$D = 0.0043 \, \vartheta T^{1.5} \tag{3-90}$$

where

$$\vartheta = \frac{\sqrt{\frac{1}{M_A} + \frac{1}{M_B}}}{(V_A^{1/3} + V_B^{1/3})} \tag{3-91}$$

where V_A and V_B is the mole volume at its boiling point of species A and B, respectively, cm^3/mol; 378 diffusion coefficients between the 27 components are obtained and are expressed by a polynomial form as

$$D = a + bT + cT^2 + dT^3 + eT^4 \tag{3-92}$$

The constants of *a, b, c* and *d* are given in Table 3-5.

Table 3.5. Constants of *a, b, c* and *d* for diffusion coefficients between different species

Species	a	b	c	d	e
C_3H_8 O_2	-1.37e-06	2e-08	9e-11	-1e-14	2e-18
C_3H_8—OH	-2e-06	4e-08	7e-11	-2e-14	3e-18
C_3H_8—H_2O_2	-1e-06	2e-08	5e-11	-1e-14	2e-18
C_3H_8—C_3H_6	-1e-06	2e-08	3e-11	-9e-15	2e-18
C_3H_8—H	-9e-06	1e-07	3e-10	-8e-14	1e-17

Species	a	b	c	d	e
C_3H_8—C_2H_3	-1e-06	2e-08	5e-11	-1e-14	2e-18
C_3H_8—HO_2	-1e-06	2e-08	5e-11	-1e-14	2e-18
C_3H_8—C_2H_4	-4e-07	1e-08	7e-11	-5e-14	2e-17
C_3H_8—C_2H_2	-4e-07	2e-08	8e-11	-5e-14	3e-17
C_3H_8—CH_2O	-1e-06	2e-08	5e-11	-1e-14	2e-18
C_3H_8—C_2H_5	-1e-06	2e-08	4e-11	-1e-14	2e-18
C_3H_8—H_2O	-1.76e-06	2e-08	1e-10	-2e-14	2e-18
C_3H_8—C_2H	-1e-06	3e-08	5e-11	-1e-14	2e-18
C_3H_8—HCO	-2e-06	3e-08	5e-11	-1e-14	2e-18
C_3H_8—CO	-1e-06	2e-08	5e-11	-1e-14	2e-18
C_3H_8—CH_3	-2e-06	3e-08	6e-11	-2e-14	3e-18
C_3H_8—C_2H_6	-1e-06	2e-08	4e-11	-1e-14	2e-18
C_3H_8—CH_2	-2e-06	3e-08	7e-11	-2e-14	3e-18
C_3H_8—CH_4	-2e-06	3e-08	6e-11	-2e-14	3e-18
C_3H_8—CO_2	-1.1e-06	1e-08	7e-11	-1e-14	1e-18
C_3H_8—O	-2e-06	4e-08	8e-11	-2e-14	4e-18
C_3H_8—C_3H_7	-1e-06	2e-08	3e-11	-9e-15	2e-18
C_3H_8—H_2	-5e-06	9e-08	2e-10	-5e-14	9e-18
C_3H_8—CH_3O	-1e-06	2e-08	4e-11	-1e-14	2e-18
C_3H_8—H_7C_3	-1e-06	2e-08	3e-11	-9e-15	2e-18
C_3H_8—C_3H_5	-1e-06	2e-08	4e-11	-1e-14	2e-18
C_3H_8—N_2	-1e-06	2e-08	9e-11	-1e-14	2e-18
O_2—OH	-4e-06	7e-08	1e-10	-4e-14	6e-18
O_2—H_2O_2	-2e-06	4e-08	7e-11	-2e-14	3e-18
O_2—C_3H_6	-2e-06	3e-08	5e-11	-1e-14	3e-18
O_2—H	-1e-05	2e-07	5e-10	-1e-13	2e-17
O_2—C_2H_3	-2e-06	4e-08	7e-11	-2e-14	3e-18
O_2—HO_2	-2e-06	4e-08	8e-11	-2e-14	4e-18
O_2—C_2H_4	-2e-06	3e-08	5e-11	-1e-14	2e-18
O_2—C_2H_2	-2e-06	4e-08	7e-11	-2e-14	3e-18
O_2—CH_2O	-2e-06	4e-08	7e-11	-2e-14	3e-18
O_2—C_2H_5	-2e-06	3e-08	7e-11	-2e-14	3e-18
O_2—H_2O	-3.14e-06	4e-08	2e-10	-3e-14	4e-18
O_2—C_2H	-2e-06	4e-08	8e-11	-2e-14	3e-18
O_2—HCO	-2e-06	4e-08	8e-11	-2e-14	4e-18
O_2—CO	-2e-06	4e-08	8e-11	-2e-14	4e-18
O_2—CH_3	-3e-06	5e-08	1e-10	-3e-14	5e-18
O_2—C_2H_6	-2e-06	3e-08	6e-11	-2e-14	3e-18
O_2—CH_2	-3e-06	5e-08	1e-10	-3e-14	5e-18
O_2—CH_4	-3e-06	5e-08	9e-11	-3e-14	4e-18
O_2—CO_2	-2e-06	2e-08	1e-10	-2e-14	3e-18
O_2—O	-4e-06	7e-08	1e-10	-4e-14	6e-18
O_2—C_3H_7	-2e-06	3e-08	5e-11	-1e-14	2e-18
O_2—H_2	-8e-06	1e-07	3e-10	-8e-14	1e-17
O_2—CH_3O	-2e-06	4e-08	7e-11	-2e-14	3e-18
O_2—H_7C_3	-2e-06	3e-08	5e-11	-1e-14	2e-18
O_2—C_3H_5	-2e-06	3e-08	5e-11	-1e-14	3e-18
O_2—N_2	-2e-06	3e-08	2e-10	-2e-14	3e-18

Table 3.5. (Continued)

Species	a	b	c	d	e
OH—H_2O_2	-3e-06	5e-08	1e-10	-3e-14	5e-18
OH—C_3H_6	-2e-06	4e-08	7e-11	-2e-14	3e-18
OH—H	-2e-05	3e-07	6e-10	-2e-13	3e-17
OH—C_2H_3	-3e-06	5e-08	9e-11	-3e-14	5e-18
OH—HO_2	-3e-06	4e-08	9e-11	-2e-14	4e-18
OH—C_2H_4	-3e-06	5e-08	9e-11	-2e-14	4e-18
OH—C_2H_2	-3e-06	5e-08	1e-10	-3e-14	5e-18
OH—CH_2O	-3e-06	5e-08	1e-10	-3e-14	5e-18
OH—C_2H_5	-3e-06	4e-08	9e-11	-2e-14	4e-18
OH—H_2O	-4e-06	7e-08	1e-10	-4e-14	6e-18
OH—C_2H	-3e-06	5e-08	1e-10	-3e-14	5e-18
OH—HCO	-3e-06	6e-08	1e-10	-3e-14	5e-18
OH—CO	-3e-06	5e-08	1e-10	-3e-14	5e-18
OH—CH_3	-4e-06	6e-08	1e-10	-4e-14	6e-18
OH—C_2H_6	-2e-06	4e-08	9e-11	-2e-14	4e-18
OH—CH_2	-4e-06	7e-08	1e-10	-4e-14	7e-18
OH—CH_4	-4e-06	6e-08	1e-10	-3e-14	6e-18
OH—CO_2	-3e-06	5e-08	9e-11	-3e-14	4e-18
OH—O	-5e-06	9e-08	2e-10	-5e-14	8e-18
OH—C_3H_7	-1e-06	2e-08	5e-11	-1e-14	2e-18
OH—H_2	-1e-05	2e-07	3e-10	-9e-14	1e-17
OH—CH_3O	-3e-06	5e-08	9e-11	-3e-14	4e-18
OH—H_7C_3	-1e-06	2e-08	5e-11	-1e-14	2e-18
OH—C_3H_5	-2e-06	4e-08	7e-11	-2e-14	4e-18
OH—N_2	-3e-06	5e-08	1e-10	-3e-14	5e-18
H_2O_2—C_3H_6	-1e-06	2e-08	5e-11	-1e-14	2e-18
H_2O_2—H	-1e-05	2e-07	4e-10	-1e-13	2e-17
H_2O_2—C_2H_3	-2e-06	3e-08	6e-11	-2e-14	3e-18
H_2O_2—HO_2	-2e-06	4e-08	7e-11	-2e-14	3e-18
H_2O_2—C_2H_4	-2e-06	3e-18	5e-11	-1e-14	2e-18
H_2O_2—C_2H_2	-2e-06	4e-08	7e-11	-2e-14	3e-18
H_2O_2—CH_2O	-2e-06	3e-08	7e-11	-2e-14	3e-18
H_2O_2—C_2H_5	-2e-06	3e-08	6e-11	-2e-14	3e-18
H_2O_2—H_2O	-3e-06	5e-08	9e-11	-3e-14	4e-18
H_2O_2—C_2H	-2e-06	4e-08	7e-11	-2e-14	3e-18
H_2O_2—HCO	-2e-06	4e-08	7e-11	-2e-14	4e-18
H_2O_2—CO	-2e-06	4e-08	7e-11	-2e-14	3e-18
H_2O_2—CH_3	-3e-06	5e-08	9e-11	-2e-14	4e-18
H_2O_2—C_2H_6	-2e-06	3e-08	6e-11	-2e-14	3e-18
H_2O_2—CH_2	-3e-06	5e-08	1e-11	-3e-14	5e-18
H_2O_2—CH_4	-3e-06	4e-08	9e-11	-2e-14	4e-18
H_2O_2—CO_2	-2e-06	3e-08	6e-11	-2e-14	3e-18
H_2O_2—O	-3e-06	6e-08	1e-10	-3e-14	6e-18
H_2O_2—C_3H_7	-1e-06	2e-08	5e-11	-1e-14	2e-18
H_2O_2—H_2	-7e-06	1e-07	3e-10	-7e-14	1e-17
H_2O_2—CH_3O	-2e-06	3e-08	6e-11	-2e-14	3e-18
H_2O_2—H_7C_3	-1e-06	2e-08	5e-11	-1e-14	2e-18
H_2O_2—C_3H_5	-1e-06	3e-08	5e-11	-1e-14	2e-18

Species	a	b	c	d	e
H_2O_2—N_2	-2e-06	4e-08	7e-11	-2e-14	3e-18
C_3H_6—H	-9e-06	1e-07	3e-10	-8e-14	1e-17
C_3H_6—C_2H_3	-1e-06	2e-08	5e-11	-1e-14	2e-18
C_3H_6—HO_2	-2e-06	3e-08	5e-11	-1e-14	3e-18
C_3H_6—C_2H_4	-1e-06	3e-08	7e-11	-2e-14	2e-18
C_3H_6—C_2H_2	-1e-06	3e-08	5e-11	-1e-14	2e-18
C_3H_6—CH_2O	-1e-06	2e-08	5e-11	-1e-14	2e-18
C_3H_6—C_2H_5	-1e-06	2e-08	7e-11	-2e-14	3e-18
C_3H_6—H_2O	-2e-06	3e-08	7e-11	-2e-14	3e-18
C_3H_6—C_2H	-2e-06	3e-08	5e-11	-1e-14	2e-18
C_3H_6—HCO	-2e-06	3e-08	5e-11	-1e-14	3e-18
C_3H_6—CO	-2e-06	3e-08	5e-11	-1e-14	2e-18
C_3H_6—CH_3	-2e-06	3e-08	7e-11	-2e-14	3e-18
C_3H_6—C_2H_6	-1e-06	2e-08	4e-11	-1e-14	2e-18
C_3H_6—CH_2	-2e-06	4e-08	7e-11	-2e-14	3e-18
C_3H_6—CH_4	-2e-06	3e-08	6e-11	-2e-14	3e-18
C_3H_6—CO_2	-1e-06	2e-08	4e-11	-1e-14	2e-18
C_3H_6—O	-2e-06	4e-08	8e-11	-2e-14	4e-18
C_3H_6—C_3H_7	-1e-06	2e-08	4e-11	-1e-14	2e-18
C_3H_6—H_2	-6e-06	1e-07	2e-10	-5e-14	9e-18
C_3H_6—CH_3O	-1e-06	2e-08	5e-11	-1e-14	2e-18
C_3H_6—H_7C_3	-1e-06	2e-08	4e-11	-1e-14	2e-18
C_3H_6—C_3H_5	-1e-06	2e-08	4e-11	-1e-14	2e-18
C_3H_6—N_2	-2e-06	3e-08	5e-11	-1e-14	2e-18
H—C_2H_3	-1e-05	2e-07	4e-10	-1e-13	2e-17
H—HO_2	-1e-05	2e-07	5e-10	-1e-13	2e-17
H—C_2H_4	-1e-05	2e-07	4e-10	-1e-13	2e-17
H—C_2H_2	-1e-05	2e-07	4e-10	-1e-13	2e-17
H—CH_2O	-1e-05	2e-07	4e-10	-1e-13	2e-17
H—C_2H_5	-1e-05	2e-07	4e-10	-1e-13	2e-17
H—H_2O	-2e-05	3e-07	5e-10	-1e-13	3e-17
H—C_2H	-1e-05	2e-07	4e-10	-1e-13	2e-17
H—HCO	-1e-05	2e-07	5e-10	-1e-13	2e-17
H—CO	-1e-05	2e-07	4e-10	-1e-13	2e-17
H—CH_3	-1e-05	2e-07	5e-10	-1e-13	2e-17
H—C_2H_6	-1e-05	2e-07	3e-10	-1e-13	2e-17
H—CH_2	-1e-05	3e-07	5e-10	-1e-13	2e-17
H—CH_4	-1e-05	2e-07	4e-10	-1e-13	2e-17
H—CO_2	-1e-05	2e-07	4e-10	-1e-13	2e-17
H—O	-2e-05	3e-07	7e-10	2e-13	3e-17
H—C_3H_7	-9e-06	2e-07	3e-10	-8e-14	1e-17
H—H_2	-2e-05	4e-07	7e-10	-2e-13	3e-17
H—CH_3O	-1e-05	2e-07	4e-10	-1e-13	2e-17
H—H_7C_3	-9e-06	2e-07	3e-10	-8e-14	1e-17
H—C_3H_5	-9e-06	2e-07	3e-10	-9e-14	2e-17
H—N_2	-1e-05	2e-07	4e-10	-1e-13	2e-17
C_2H_3—HO_2	-2e-06	4e-08	7e-11	-2e-14	3e-18
C_2H_3—C_2H_4	-2e-06	3e-08	6e-11	-2e-14	3e-18

Table 3.5. (Continued)

Species	a	b	c	d	e
C_2H_3—C_2H_2	-2e-06	3e-08	7e-11	-2e-14	3e-18
C_2H_3—CH_2O	-2e-06	3e-08	6e-11	-2e-14	3e-18
C_2H_3—C_2H_5	-2e-06	3e-08	6e-11	-2e-14	3e-18
C_2H_3—H_2O	-3e-06	4e-08	9e-11	-3e-14	5e-18
C_2H_3—C_2H	-2e-06	3e-08	7e-11	-2e-14	3e-18
C_2H_3—HCO	-2e-06	4e-08	7e-11	-2e-14	3e-18
C_2H_3—CO	-2e-06	3e-08	7e-11	-2e-14	3e-18
C_2H_3—CH_3	-3e-06	4e-08	9e-11	-2e-14	4e-18
C_2H_3—C_2H_6	-2e-06	3e-08	6e-11	-2e-14	3e-18
C_2H_3—CH_2	-3e-06	5e-08	9e-11	-3e-14	4e-18
C_2H_3—CH_4	-2e-06	4e-08	8e-11	-2e-14	4e-18
C_2H_3—CO_2	-2e-06	3e-08	6e-11	-2e-14	3e-18
C_2H_3—O	-3e-06	6e-08	1e-10	-3e-14	5e-18
C_2H_3—C_3H_7	-1e-06	2e-08	5e-11	-1e-14	3e-18
C_2H_3—H_2	-7e-06	1e-07	2e-10	-7e-14	1e-17
C_2H_3—CH_3O	-2e-06	3e-08	6e-11	-2e-14	3e-18
C_2H_3—H_7C_3	-1e-06	2e-08	5e-11	-1e-14	3e-18
C_2H_3—C_3H_5	-1e-06	2e-08	5e-11	-1e-14	2e-18
C_2H_3—N_2	-3e-06	5e-08	9e-11	-3e-14	5e-18
HO_2—C_2H_4	-2e-06	3e-08	7e-11	-2e-14	3e-18
HO_2—C_2H_2	-2e-06	4e-08	7e-11	-2e-14	4e-18
HO_2—CH_2O	-2e-06	3e-08	6e-11	-2e-14	3e-18
HO_2—C_2H_5	-2e-06	3e-08	5e-11	-1e-14	3e-18
HO_2—H_2O	-3e-06	5e-08	1e-10	-3e-14	5e-18
HO_2—C_2H	-2e-06	4e-08	8e-11	-2e-14	4e-18
HO_2—HCO	-2e-06	4e-08	8e-11	-2e-14	4e-18
HO_2—CO	-2e-06	4e-08	8e-11	-2e-14	4e-18
HO_2—CH_3	-3e-06	5e-08	1e-10	-3e-14	5e-18
HO_2—C_2H_6	-2e-06	3e-08	6e-11	-2e-14	3e-18
HO_2—CH_2	-3e-06	5e-08	1e-10	-3e-14	5e-18
HO_2—CH_4	-3e-06	5e-08	9e-11	-3e-14	4e-18
HO_2—CO_2	-2e-06	3e-08	7e-11	-2e-14	3e-18
HO_2—O	-4e-06	6e-08	1e-10	-3e-14	6e-18
HO_2—C_3H_7	-2e-06	3e-08	5e-11	-1e-14	2e-18
HO_2—H_2	-8e-06	1e-07	3e-10	-8e-14	1e-17
HO_2—CH_3O	-2e-06	4e-08	7e-11	-2e-14	3e-18
HO_2—H_7C_3	-2e-06	3e-08	5e-11	-1e-14	2e-18
HO_2—C_3H_5	-2e-06	3e-08	5e-11	-1e-14	3e-18
HO_2—N_2	-2e-06	3e-08	6e-11	-2e-14	3e-18
C_2H_4—C_2H_2	-2e-06	3e-08	6e-11	-2e-14	3e-18
C_2H_4—CH_2O	-2e-06	3e-08	6e-11	-2e-14	3e-18
C_2H_4—C_2H_5	-2e-06	3e-08	9e-11	-2e-14	4e-18
C_2H_4—H_2O	-3e-06	4e-08	9e-11	-2e-14	4e-18
C_2H_4—C_2H	-2e-06	3e-08	7e-11	-2e-14	3e-18
C_2H_4—HCO	-2e-06	3e-08	7e-11	-2e-14	3e-18
C_2H_4—CO	-2e-06	3e-08	7e-11	-2e-14	3e-18
C_2H_4—CH_3	-2e-06	4e-08	8e-11	-2e-14	4e-18
C_2H_4—C_2H_6	-2e-06	3e-08	5e-11	-1e-14	3e-18

Species	a	b	c	d	e
C_2H_4—CH_2	-3e-06	4e-08	9e-11	-2e-14	4e-18
C_2H_4—CH_4	-2e-06	4e-08	8e-11	-2e-14	4e-18
C_2H_4—CO_2	-2e-06	3e-08	6e-11	-2e-14	3e-18
C_2H_4—O	-3e-06	5e-08	1e-10	-3e-14	5e-18
C_2H_4—C_3H_7	-1e-06	2e-08	5e-11	-1e-14	2e-18
C_2H_4—H_2	-7e-06	1e-07	2e-10	-6e-14	1e-17
C_2H_4—CH_3O	-2e-06	3e-08	6e-11	-2e-14	3e-18
C_2H_4—H_7C_3	-1e-06	2e-08	5e-11	-1e-14	2e-18
C_2H_4—C_3H_5	-1e-06	2e-08	5e-11	-1e-14	2e-18
C_2H_4—N_2	-2e-06	3e-08	6e-11	-2e-14	3e-18
C_2H_2—CH_2O	-2e-06	3e-08	7e-11	-2e-14	3e-18
C_2H_2—C_2H_5	-2e-06	3e-08	6e-11	-2e-14	3e-18
C_2H_2—H_2O	-3e-06	5e-08	9e-11	-3e-14	4e-18
C_2H_2—C_2H	-2e-06	4e-08	7e-11	-2e-14	3e-18
C_2H_2—HCO	-2e-06	4e-08	7e-11	-2e-14	4e-18
C_2H_2—CO	-2e-06	3e-08	7e-11	-2e-14	3e-18
C_2H_2—CH_3	-3e-06	5e-08	9e-11	-2e-14	4e-18
C_2H_2—C_2H_6	-2e-06	3e-08	6e-11	-2e-14	3e-18
C_2H_2—CH_2	-3e-06	5e-08	1e-10	-3e-14	5e-18
C_2H_2—CH_4	-2e-06	4e-08	8e-11	-2e-14	4e-18
C_2H_2—CO_2	-2e-06	3e-08	6e-11	-2e-14	3e-18
C_2H_2—O	-3e-06	6e-08	1e-10	-3e-14	5e-18
C_2H_2—C_3H_7	-1e-06	2e-08	5e-11	-1e-14	2e-18
C_2H_2—H_2	-7e-06	1e-07	2e-10	-7e-14	1e-17
C_2H_2—CH_3O	-2e-06	3e-08	6e-11	-2e-14	3e-18
C_2H_2—H_7C_3	-1e-06	2e-08	5e-11	-1e-14	2e-18
C_2H_2—C_3H_5	-2e-06	3e-08	5e-11	-1e-14	2e-18
C_2H_2—N_2	-2e-06	4e-08	7e-11	-2e-14	3e-18
CH_2O—C_2H_5	-2e-06	3e-08	6e-11	-2e-14	3e-18
CH_2O—H_2O	-3e-06	4e-08	8e-11	-2e-14	4e-18
CH_2O—C_2H	-2e-06	3e-08	7e-11	-2e-10	3e-18
CH_2O—HCO	-2e-06	4e-08	7e-11	-2e-14	3e-18
CH_2O—CO	-2e-06	3e-08	7e-11	-2e-14	3e-18
CH_2O—CH_3	-3e-06	4e-08	9e-11	-2e-14	4e-18
CH_2O—C_2H_6	-2e-06	3e-08	6e-11	-2e-14	3e-18
CH_2O—CH_2	-3e-06	5e-08	9e-11	-3e-14	4e-18
CH_2O—CH_4	-2e-06	4e-08	8e-11	-2e-14	4e-18
CH_2O—CO_2	-2e-06	4e-08	8e-11	-2e-14	4e-18
CH_2O—O	-3e-06	6e-08	1e-10	-3e-14	6e-18
CH_2O—C_3H_7	-1e-06	2e-08	5e-11	-1e-14	2e-18
CH_2O—H_2	-7e-06	1e-07	2e-10	-7e-14	1e-17
CH_2O-CH_3O	-2e-06	3e-08	6e-11	-2e-14	3e-18
CH_2O—H_7C_3	-1e-06	2e-08	5e-11	-1e-14	2e-18
CH_2O—C_3H_5	-1e-06	2e-08	5e-11	-1e-14	2e-18
CH_2O—N_2	-2e-06	3e-08	7e-11	-2e-14	3e-18
C_2H_5—H_2O	-2e-06	4e-08	8e-11	-2e-14	4e-18
C_2H_5—C_2H	-2e-06	3e-08	7e-11	-2e-10	3e-18
C_2H_5—HCO	-2e-06	3e-08	7e-11	-2e-14	3e-18

Table 3.5. (Continued)

Species	a	b	c	d	e
C_2H_5—CO	-2e-06	3e-08	6e-11	-2e-14	3e-18
C_2H_5—CH_3	-2e-06	4e-08	8e-11	-2e-14	4e-18
C_2H_5—C_2H_6	-2e-06	3e-08	5e-11	-1e-14	3e-18
C_2H_5—CH_2	-3e-06	4e-08	9e-11	-2e-14	4e-18
C_2H_5—CH_4	-2e-06	4e-08	8e-11	-2e-14	4e-18
C_2H_5—CO_2	-2e-06	3e-08	5e-11	-2e-14	3e-18
C_2H_5—O	-3e-06	5e-08	1e-10	-3e-14	5e-18
C_2H_5—C_3H_7	-1e-06	2e-08	4e-11	-1e-14	2e-18
C_2H_5—H_2	-6e-06	1e-07	2e-10	-6e-14	1e-17
C_2H_5—CH_3O	-2e-06	3e-08	6e-11	-2e-14	3e-18
C_2H_5—H_7C_3	-1e-06	2e-08	4e-11	-1e-14	2e-18
C_2H_5—C_3H_5	-1e-06	2e-08	5e-11	-1e-14	2e-18
C_2H_5—N_2	-2e-06	3e-08	6e-11	-2e-14	3e-18
H_2O—C_2H	-3e-06	5e-08	1e-10	-3e-14	5e-18
H_2O—HCO	-3e-06	5e-08	1e-10	-3e-14	5e-18
H_2O—CO	-3e-06	5e-08	1e-10	-3e-14	5e-18
H_2O—CH_3	-4e-06	6e-08	1e-10	-3e-14	-3e-14
H_2O—C_2H_6	-2e-06	4e-08	8e-11	-2e-14	4e-18
H_2O—CH_2	-4e-06	7e-08	1e-10	-4e-14	6e-18
H_2O—CH_4	-3e-06	6e-08	1e-10	-3e-14	5e-18
H_2O—CO_2	-3e-06	3e-08	2e-10	-2e-14	3e-18
H_2O—O	-5e-06	8e-08	2e-10	-4e-14	8e-18
H_2O—C_3H_7	-2e-06	3e-08	7e-11	-2e-14	3e-18
H_2O—H_2	-9e-06	2e-07	3e-10	-9e-14	1e-17
H_2O—CH_3O	-3e-06	4e-08	9e-11	-2e-14	4e-18
H_2O—H_7C_3	-2e-06	3e-08	7e-11	-2e-14	3e-18
H_2O—C_3H_5	-2e-06	4e-08	7e-11	-2e-14	3e-18
H_2O—N_2	-3.14e-06	4e-08	2e-10	-3e-14	4e-18
C_2H—HCO	-2e-06	4e-08	8e-11	-2e-14	4e-18
C_2H—CO	-2e-06	4e-08	7e-11	-2e-14	4e-18
C_2H—CH_3	-3e-06	5e-08	9e-11	-3e-14	4e-18
C_2H—C_2H_6	-2e-06	3e-08	6e-11	-2e-14	3e-18
C_2H—CH_2	-3e-06	5e-08	1e-10	-3e-14	5e-18
C_2H—CH_4	-3e-06	4e-08	9e-11	-2e-14	4e-18
C_2H—CO_2	-2e-06	3e-08	6e-11	-2e-14	3e-18
C_2H—O	-3e-06	5e-08	1e-10	-3e-14	5e-18
C_2H—C_3H_7	-1e-06	3e-08	5e-11	-1e-14	2e-18
C_2H—H_2	-3e-06	6e-08	1e-10	-3e-14	5e-18
C_2H—CH_3O	-2e-06	3e-08	7e-11	-2e-14	3e-18
C_2H—H_7C_3	-1e-06	3e-08	5e-11	-1e-14	2e-18
C_2H—C_3H_5	-2e-06	3e-08	5e-11	-1e-14	3e-18
C_2H—N_2	-2e-06	4e-08	7e-11	-2e-14	4e-18
HCO—CO	-2e-06	4e-08	8e-11	-2e-14	4e-18
HCO—CH_3	-3e-06	5e-08	1e-10	-3e-14	4e-18
HCO—C_2H_6	-2e-06	3e-08	6e-11	-2e-14	3e-18
HCO—CH_2	-3e-06	5e-08	1e-10	-3e-14	5e-18
HCO—CH_4	-3e-06	5e-08	9e-11	-3e-14	4e-18
HCO—CO_2	-2e-06	3e-08	7e-11	-2e-14	3e-18

Species	a	b	c	d	e
HCO—O	-4e-06	7e-08	1e-10	-4e-14	6e-18
HCO—C3H_7	-2e-06	3e-08	5e-11	-1e-14	2e-18
HCO—H_2	-8e-06	1e-07	3e-10	-8e-14	1e-17
HCO—CH_3O	-2e-06	4e-08	7e-11	-2e-14	3e-18
HCO—H_7C_3	-2e-06	3e-08	5e-11	-1e-14	2e-18
HCO—C_3H_5	-2e-06	3e-08	6e-11	-2e-14	3e-18
HCO—N_2	-2e-06	4e-08	8e-11	-2e-14	4e-18
CO—CH_3	-3e-06	5e-08	1e-10	-3e-14	5e-18
CO—C_2H_6	-2e-06	3e-08	6e-11	-2e-14	3e-18
CO—CH_2	-3e-06	5e-08	1e-10	-3e-14	5e-18
CO—CH_4	-3e-06	4e-08	9e-11	-2e-14	4e-18
CO—CO_2	-2e-06	3e-08	6e-11	-2e-14	3e-18
CO—O	-4e-06	6e-08	1e-10	-3e-14	6e-18
CO—C_3H_7	-1e-06	3e-08	5e-11	-1e-14	2e-18
CO—H_2	-8e-06	1e-07	3e-10	-7e-14	1e-17
CO—CH_3O	-2e-06	3e-08	7e-11	-2e-14	3e-18
CO—H_7C_3	-1e-06	3e-08	5e-11	-1e-14	2e-18
CO—C_3H_5	-2e-06	3e-08	5e-11	-1e-14	3e-18
CO—N_2	-2e-06	4e-08	7e-11	-2e-14	4e-18
CH_3—C_2H_6	-2e-06	4e-08	8e-11	-2e-14	4e-18
CH_3—CH_2	-4e-06	6e-08	1e-10	-3e-14	6e-18
CH_3—CH_4	-3e-06	5e-08	1e-10	-3e-14	5e-18
CH_3—CO_2	-2e-06	4e-08	8e-11	-2e-14	4e-18
CH_3—O	-4e-06	8e-08	2e-10	-4e-14	7e-18
CH_3—C_3H_7	-2e-06	3e-08	7e-11	-2e-14	3e-18
CH_3—H_2	-8e-06	1e-07	3e-10	-8e-14	1e-17
CH_3—CH_3O	-3e-06	4e-08	9e-11	-2e-14	4e-18
CH_3—H_7C_3	-2e-06	3e-08	7e-11	-2e-14	3e-18
CH_3—C_3H_5	-2e-06	3e-08	7e-11	-2e-14	3e-18
CH_3—N_2	-3e-06	5e-08	9e-11	-3e-14	4e-18
C_2H_6—CH_2	-2e-06	4e-08	8e-11	-2e-14	4e-18
C_2H_6—CH_4	-2e-06	4e-08	7e-11	-2e-14	3e-18
C_2H_6—CO_2	-2e-06	3e-08	5e-11	-1e-14	3e-18
C_2H_6—O	-3e-06	5e-08	1e-10	-3e-14	5e-18
C_2H_6—C_3H_7	-1e-06	2e-08	4e-11	-1e-14	2e-18
C_2H_6—H_2	-6e-06	1e-07	2e-10	-6e-14	1e-17
C_2H_6—CH_3O	-2e-06	3e-08	6e-11	2e-14	3e-18
C_2H_6—H_7C_3	-1e-06	2e-08	4e-11	-1e-14	2e-18
C_2H_6—C_3H_5	-1e-06	2e-08	4e-11	-1e-14	2e-18
C_2H_6—N_2	-2e-06	3e-08	9e-11	-2e-14	3e-18
CH_2—CH_4	-3e-06	6e-08	1e-10	-3e-14	5e-18
CH_2—CO_2	-3e-06	5e-08	9e-11	-2e-14	4e-18
CH_2—O	-5e-06	8e-08	2e-10	-4e-14	8e-18
CH_2—C_3H_7	-2e-06	4e-08	7e-11	-2e-14	3e-18
CH_2—H_2	-9e-06	2e-07	3e-10	-8e-14	1e-17
CH_2—CH_3O	-3e-06	5e-08	9e-11	-3e-14	4e-18
CH_2—H_7C_3	-2e-06	4e-08	7e-11	-2e-14	3e-18
CH_2—C_3H_5	-2e-06	4e-08	7e-11	-2e-14	3e-18

Table 3.5. (Continued)

Species	a	b	c	d	e
CH_2—N_2	-3e-06	5e-08	1e-10	-3e-14	5e-18
CH_4—CO_2	-2e-06	4e-08	8e-11	-2e-14	4e-18
CH_4—O	-3e-06	5e-08	1e-10	-3e-14	5e-18
CH_4—C_3H_7	-2e-06	3e-08	6e-11	-2e-14	3e-18
CH_4—H_2	-8e-06	1e-07	3e-10	-7e-14	1e-17
CH_4—CH_3O	-2e-06	4e-08	8e-11	-2e-14	4e-18
CH_4—H_7C_3	-2e-06	3e-08	6e-11	-2e-14	3e-18
CH_4—C_3H_5	-2e-06	3e-08	6e-11	-2e-14	3e-18
CH_4—N_2	-3e-06	4e-08	9e-11	-2e-14	4e-18
CO_2—O	-3e-06	5e-08	1e-10	-3e-14	5e-18
CO_2—C_3H_7	-1e-06	2e-08	4e-11	-1e-14	2e-18
CO_2—H_2	-8e-06	1e-07	3e-10	-8e-14	1e-17
CO_2—CH_3O	-2e-06	3e-08	6e-11	-2e-14	3e-18
CO_2—H_7C_3	-1e-06	2e-08	4e-11	-1e-14	2e-18
CO_2—C_3H_5	-1e-06	2e-08	5e-11	-1e-14	2e-18
CO_2—N_2	-2e-06	2e-08	1e-10	-2e-14	3e-18
O—C_3H_7	-2e-06	4e-08	8e-11	-2e-14	4e-18
O—H_2	1e-05	2e-07	4e-10	1e-13	2e-17
O—CH_3O	-3e-06	6e-08	1e-10	-3e-14	5e-18
O—H_7C_3	-2e-06	4e-08	8e-11	-2e-14	4e-18
O—C_3H_5	-3e-06	4e-08	9e-11	-2e-14	4e-18
O—N_2	-5e-06	9e-08	2e-10	-5e-14	8e-18
C_3H_7—H_2	-5e-06	9e-08	2e-10	-5e-14	9e-18
C_3H_7—CH_3O	-1e-06	2e-08	5e-11	-1e-14	2e-18
C_3H_7—H_7C_3	-1e-06	2e-08	3e-11	-1e-14	2e-18
C_3H_7—C_3H_5	-1e-06	2e-08	4e-11	-1e-14	2e-18
C_3H_7—N_2	-1e-06	3e-08	5e-11	-1e-14	2e-18
H_2—CH_3O	-7e-06	1e-07	2e-10	-7e-14	1e-17
H_2—H_7C_3	-5e-06	9e-08	2e-10	-5e-14	9e-18
H_2—C_3H_5	-6e-06	1e-07	2e-10	-5e-14	9e-18
H_2—N_2	-8e-06	1e-07	3e-10	-7e-14	1e-17
CH_3O—H_7C_3	-1e-06	2e-08	5e-11	-1e-14	2e-18
CH_3O—C_3H_5	-1e-06	2e-08	5e-11	-1e-14	2e-18
CH_3O—N_2	-2e-06	3e-08	7e-11	-2e-14	3e-18
H_7C_3—C_3H_5	-1e-06	2e-08	4e-11	-1e-14	2e-18
H_7C_3—N_2	-1e-06	3e-08	5e-11	-1e-14	2e-18
C_3H_5—N_2	-2e-06	3e-08	5e-11	-1e-14	3e-18

The source term, in general, is not a constant, but depends on the unknown variable φ, and local linearization in a control volume P is made as

$$S = S_c + S_p \varphi_p \tag{3-93}$$

where S_c is a constant and is calculated by the φ value of the last iteration from S *versus* φ function, S_p is the slop of S *versus* φ curve at the point P and linearization has to be such made that S_p is negative for the stability in the simulations.

In the simulations, the turbulence is simulated by Standard k-ε model and controlled by non-equilibrium wall functions in the wall region. SIMPLE method is selected for pressure-velocity coupling and the convection term is discretized with fist-order upwind scheme. Radiant energy absorption and irradiant are incorporated in the enthalpy balance by P1 model.

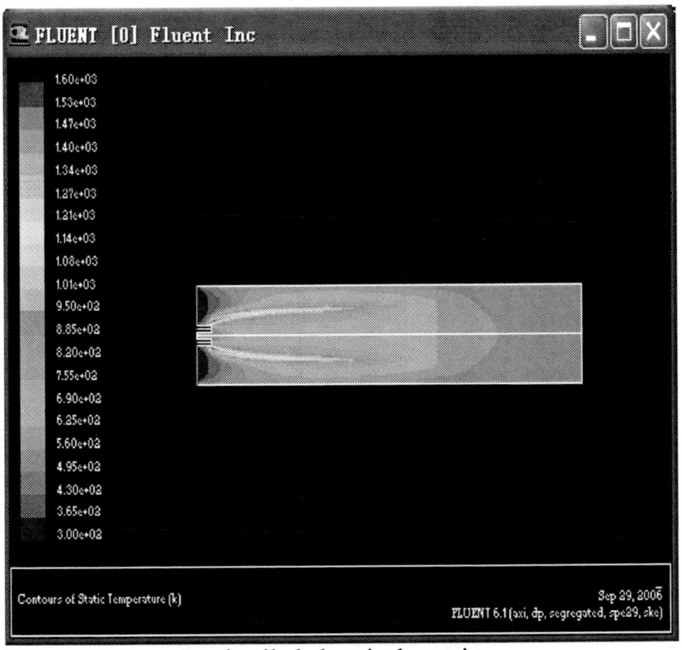

By detailed chemical reactions

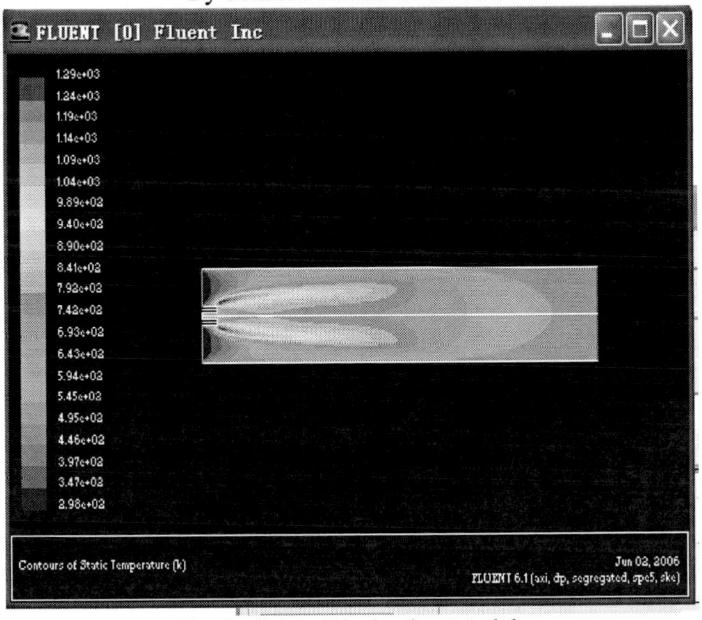

By the Eddy-Dissipation Model.

Figure 3.6. Simulated temperature contour plot.

3.2.5.4. Simulation Results

Simulated temperature contour plot for the test given in Table 3-2 is shown in Figure 3-6. The simulation results from the combustion rates controlled by the Eddy-Dissipation Model (EDM) are also presented in Figure 3-6. The figure shows that two methods give similar predictions on temperature profiles. The maximum temperature is about 1100K and occurs in the mid of central axis and the wall, which is slightly lower than the measured value. Figure 3-7 shows mass fractions of propane, O_2, CO_2 and H_2O along the central axis predicted by detailed chemical reactions and EDM method. It can be seen that the two methods give close predictions and the flame length where propane concentration declined to zero is about 0.35m, which is consistent with that observed visually.

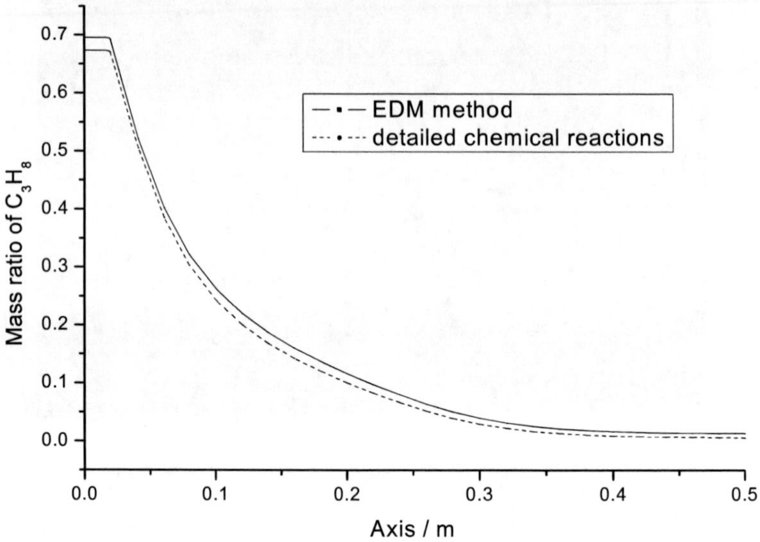

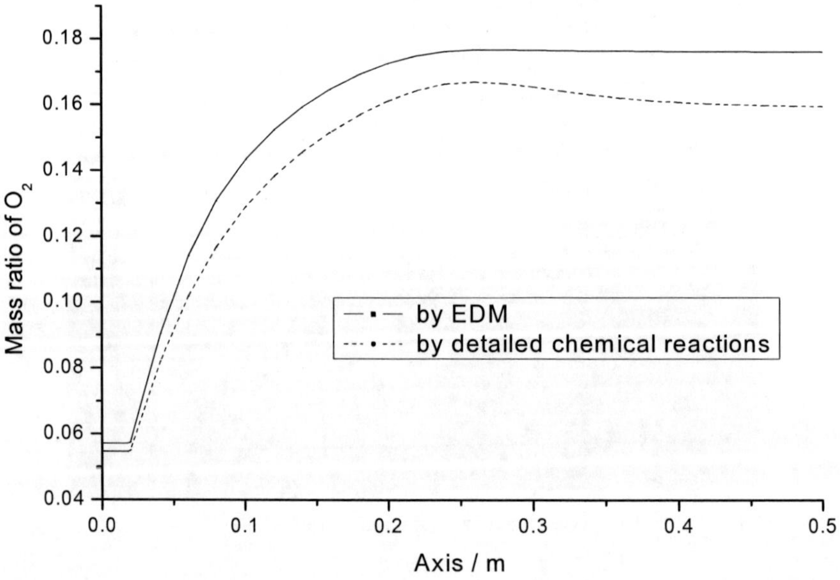

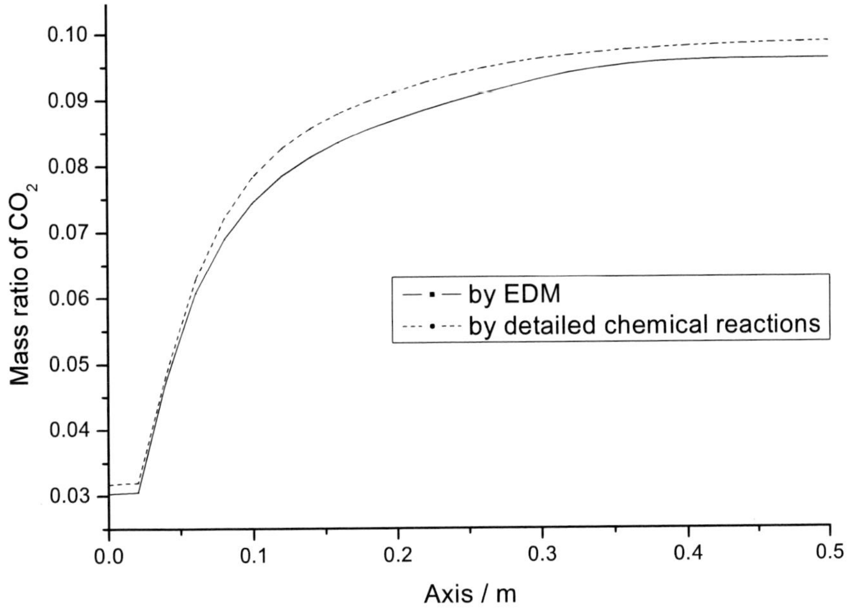

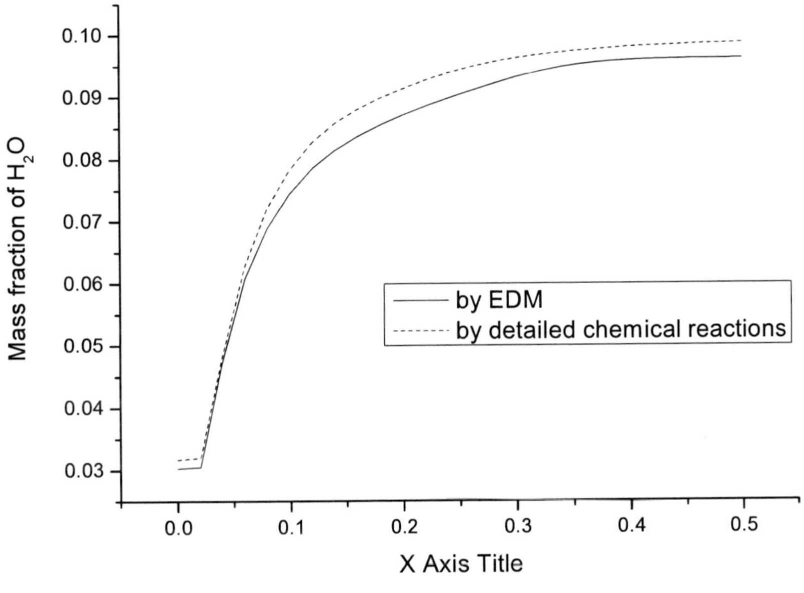

Figure 3.7. Simulated C_3H_8, O_2, CO_2 and H_2O profiles along central axis.

Mass fractions of reaction components and intermediates are grouped according to their orders of magnitude. The first group contains C_2H, C_2H_4, C_2H_6, C_3H_5, C_3H_6, and H_2 and their mass fraction is in an order of magnitude of 10^{-3} inside the flame (the flame length is about 0.35m as simulated and 0.335m as observed), as shown in Figure 3-8. The second group contains H_7C_3, CH_2O, C_2H_5, H_2O_2, and CO and their mass fraction is in an order of magnitude of 10^{-5} outside the flame except CO, as shown in Figure 3-9. The third group contains C_3H_7, C_2H_3, and CH_4 and their mass fraction is in an order of magnitude of 10^{-7} with CH_4 inside and C_3H_7, C_2H_3 outside the flame, as shown in Figure 3-10. The mass fraction of all other reaction components and intermediates is in much lower orders of magnitudes, except C_2H_2.

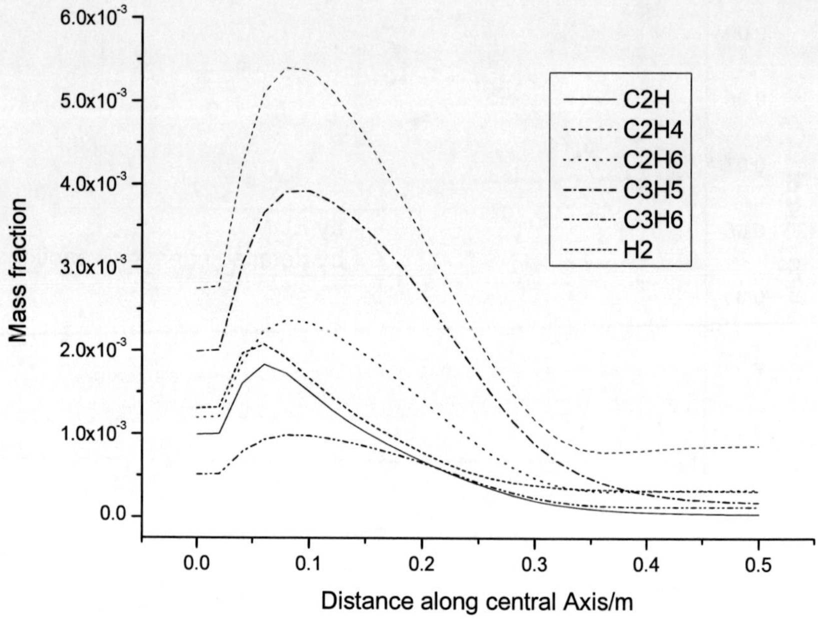

Figure 3.8. Simulated mass fractions of C_2H, C_2H_4, C_2H_6, C_3H_5, C_3H_6, and H_2 intermediates along central axis.

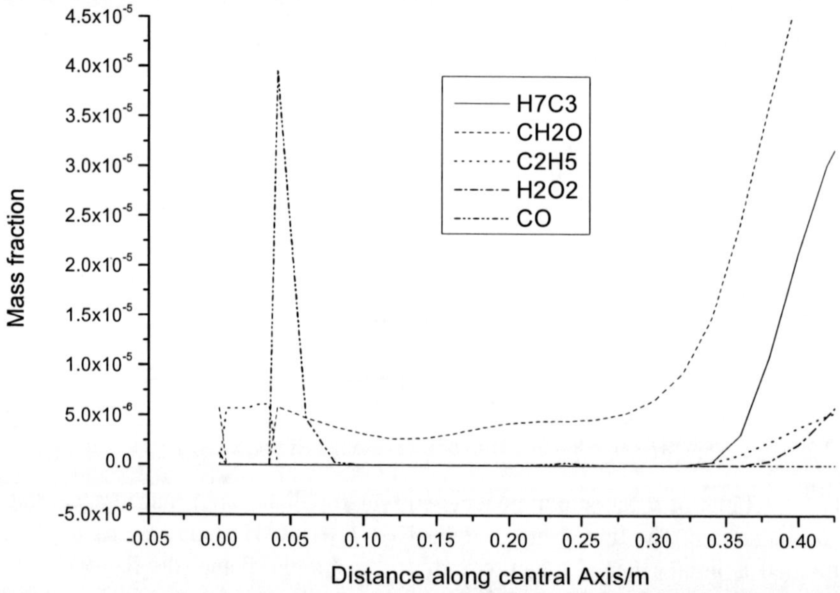

Figure 3.9. Simulated mass fractions of H_7C_3, CH_2O, C_2H_5, H_2O_2, and CO intermediates along central axis.

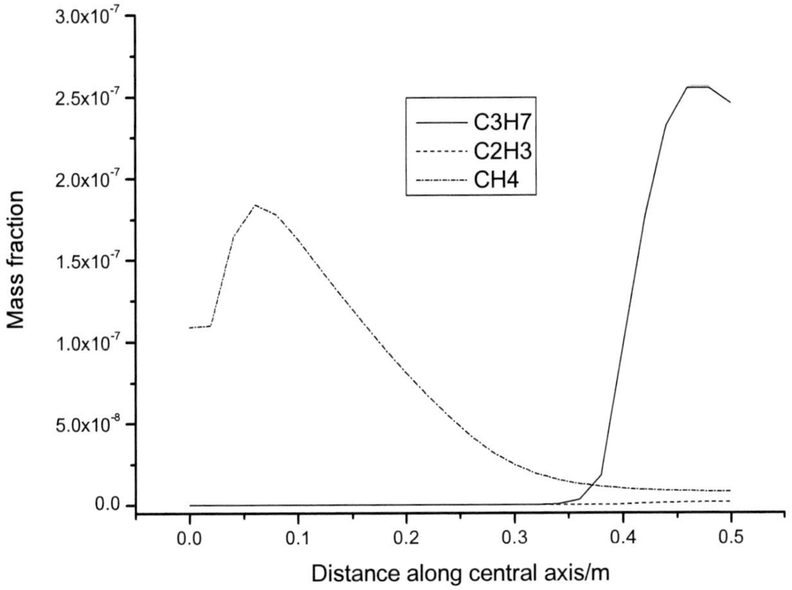

Figure 3.10. Simulated mass fractions of C_3H_7, C_2H_3, and CH_4 intermediates along central axis.

Simulated mass fraction of C_2H_2 is compared with experimental data in Figure 3-11. It can be seen that mass fraction of C_2H_2 is in the order of magnitude of 10^{-2}, and detailed chemical reaction method give reasonable predictions on C_2H_2 concentration. Simulations for other three test runs are also presented in Figure 3-11. Operational conditions and measured results of the three test runs are listed in Table 3-6.

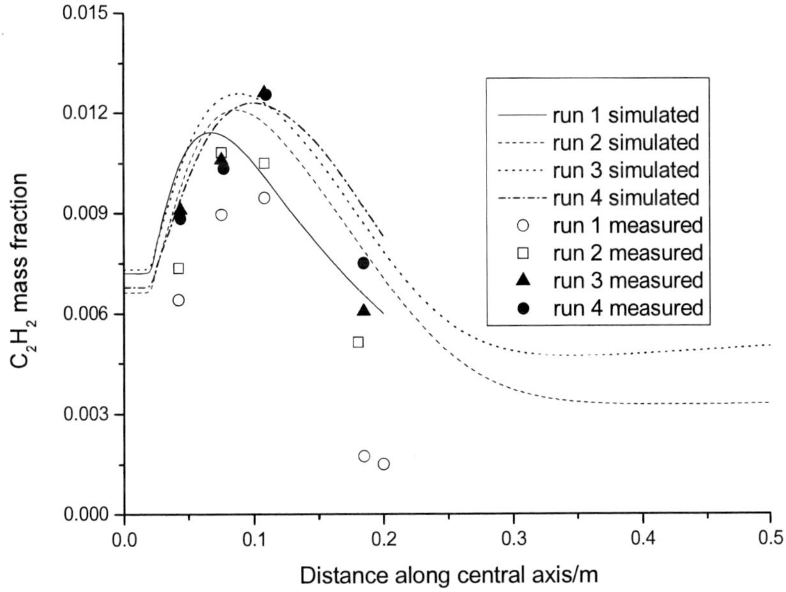

Figure 3.11. C_2H_2 mass fraction between simulated and measured.

Table 3.6. Operational conditions and measurement results

Test Flame	1	2	3	4
Pressure (atm)	1.05	1.05	1.05	1.05
C_3H_8 flow rate (mol/s)	0.000777	0.001033	0.001292	0.00155
Air flow rate (mol/s)	0.015133	0.015133	0.015133	0.015133
Flame length (m)	0.335	0.335	0.335	0.445
Flame temperature(K)	1150	1173	1220	1265
Particle size by TEM (nm)	28	33.96	36.3	30.77
Soot yield (g/min)	0.42	1.56	2.28	1.448
Re (-)[a]	939.1	1215.6	1564.2	1875.2

[a] Nominal average value based on an injection temperature of 298K.

3.2.6. Simulation of Soot Yield

The earliest stage of soot generation is believed to be the formation of 5-10 nm particles due to reactive coagulation of polycyclic aromatic hydrocarbon (PAH) molecules [3-8]. These particles are more transparent to a transmission electron microscope (TEM) electron beam than the more mature soot particles and are often referred as soot precursor particles [5,9,10]. The later stages are characterized by simultaneous coagulation, coalescence and surface growth of the precursors leading to an increase of the particle size. Simultaneously an internal rearrangement of the precursor particles occurs resulting in structures with more condensed aromatic rings and more compact shape called "Soot nuclei" are observed by HRTEM and SAXS. These nuclei grow due to surface reactions and aggregation which is a hierarchical process i.e., the well known fractal-like aggregates or globular agglomerates forming by clustering of primary particles, forming the final aggregates.

Study of the process is of significant since the carbon black (CB) particles in the atmosphere badly affect the air quality and human health. A survey of epidemic shows the high relevancy of the particles with diameter less than 10μm and the increasing death rate, recent studies discovered that the ultra fine particles whose diameters are less than 0.1μm are even more harmful to the human respiratory system. Generally speaking, about 50% of the fine particles in the air are carbon compound, which has the forms of organic carbon (organic compound and functional group) and element carbon (carbon black). The reflectivity of soot aerosols determines the length of sun radiation time and the organic carbon can absorb the infrared, which results in the global warming. Since the combustion process is the main source of the carbon particles, understanding of the soot formation in combustion is important to reduce the production and thus to reduce the harm to human body as well as maintain the radiation balance.

In the simulation of soot growth, both surface growth rate and nucleation rate are associated with intermediates such as acetylene concentration [11-13], and other soot precursor radicals [14-15]. But in either case, detailed chemical reaction mechanisms have to be simulated and the simulation is cost prohibitive and quantitative agreement with experimental data remains a challenge.

3.2.6.1. Acetylene Concentration

Acetylene concentration is measured by gas sampling from tapings on combustor wall as shown in Figure 3-12. Four tapings at different height levels allow to measure acetylene concentration along the combustor height. The combustion chamber is a stainless tube with outer diameter a 110mm and 3.5mm wall thickness, and 0.5m height. A fine sieve of 600mesh is placed at the top of the combustion chamber to avoid the soot particles emitted into the ambiance and to collect the soot particles for TEM measurements. The burner is made up of two concentric stainless-steel tubes (8×1 and 16×1 in their nominal size and wall thickness). Air flows into the combustion chamber through a perforated plate in the annular space between the burner and the wall of the combustion chamber.

Gaseous samples for analysis are taken through the tapings on the chamber wall by sucking from the central axis of the chamber with a 0.8 mm stainless steel capillary pipe (sucking rate was about 2ml/s). Gas chromatography (GC-14C GAS CHROMATOGRAPH) is used to analysis gas compositions and their concentrations. Identification of species is accomplished by matching the gas-chromatographic retention times to pure components and is given in Table 3-7.

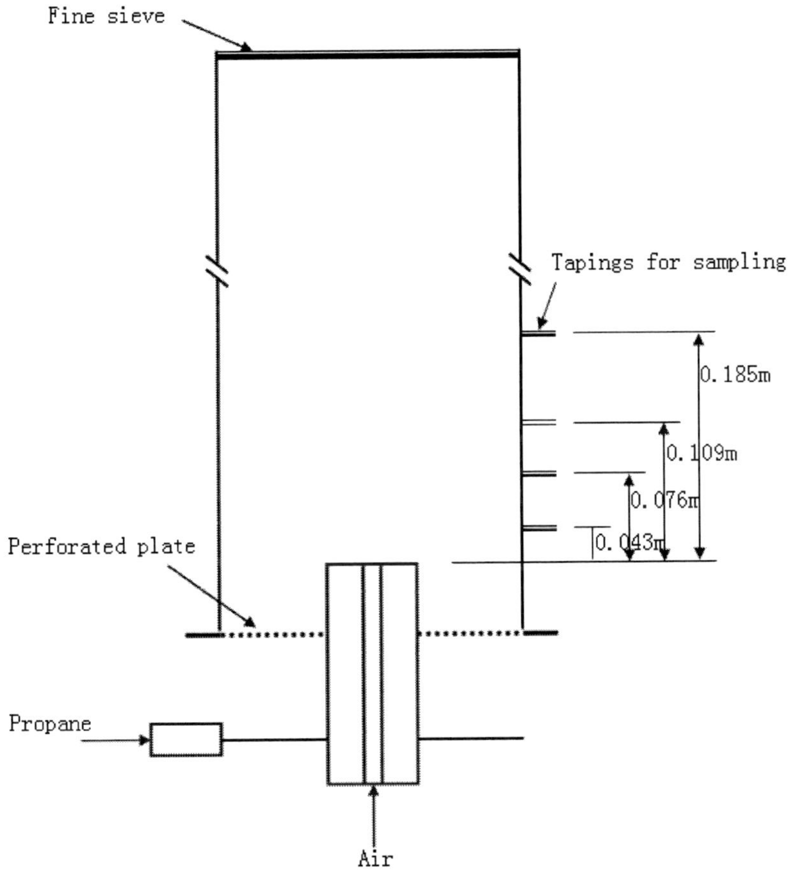

Figure 3.12. Schematic diagram of the experimental setup for acetylene and soot particles.

Table 3.7. Standard parameters of gas chromatograph

standard gas	retention time	Peak area	Mole ratio
CH_4	1.5	415,000	301.2ppm
CO	3.55	393,200	293.3ppm
C_2H_2	20.6	575,600,000	100%
CO_2	7.3	393,200	293.3ppm
C_3H_8	19.441	8.44526	0.95

Capillary column 50m at temperature 80°C. FID detector at temperature 120°C.

A typical GC spectrum of gas samples is shown in Figure 3-13 for test run 2 at the height of 0.185m. From the retention times, it can be seen that CH_4, CO, CO_2 and C_3H_8 are detected. From the peak areas, mole fractions of CH_4, CO and CO_2 can be obtained. Such obtained mole fractions of CH_4, CO and CO_2 in the four flames are tabulated in Table 3-8. Mass fraction of acetylene along flame height of the four test runs is plotted in Figure 3-11, it could be seen that acetylene increases rapidly from the bottom of flame, reaches to maximum at a height level about 0.1m, and then decreases quickly from the maximum to a height about 0.2m.

Experimental results show that the concentration of acetylene in the flame is the only hydrocarbons present in significant quantities in different operation conditions. This conforms the experimental results with premixed flames of that soot growth primarily depends on the concentration of acetylene and soot formation ends when acetylene disappears [6,11-13].

From gas state equation, the acetylene concentration can be obtained as

$$[C_2H_2] = C_m \frac{273}{22.4T} = 12.1875 \frac{C_m}{T} \tag{3-94}$$

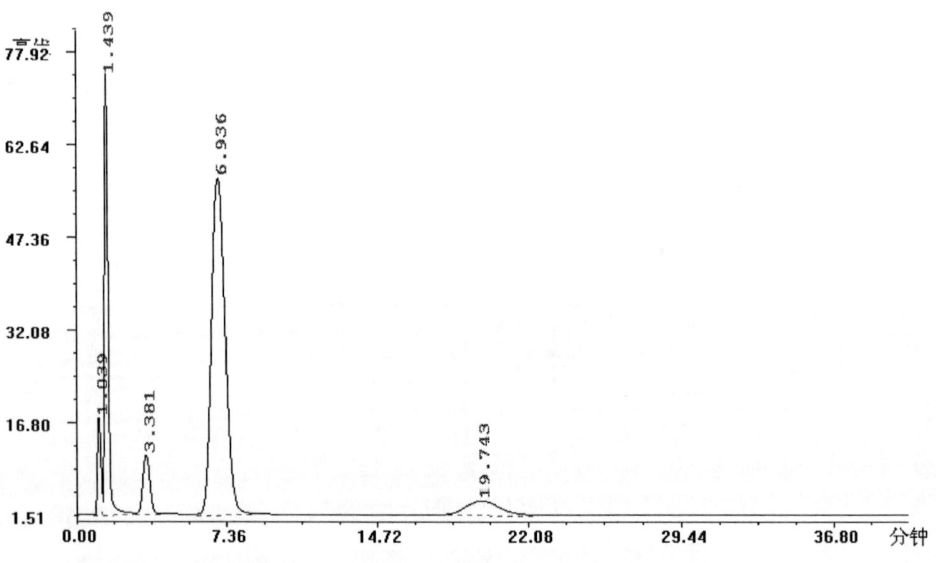

Figure 3.13. Typical GC spectrum of gas sample (test run 2).

where C_m is acetylene mole fraction. The results show that the concentration of acetylene in the flame could be fitted by the Rayleigh distribution function along the axis of the flame as

$$[C_2H_2] = ax\exp(-bx^2) \tag{3-95}$$

as shown in Figure 3-14. The constants *a* and *b* are tabulated in Table 3-9.

Table 3.8. Mole fraction of gases along flame axis under various conditions

Flame	Height (m)	CH_4 mole fraction	CO mole fraction	CO_2 mole fraction	C_2H_2 mole fraction
Run 1	0.043	0.0045956	0.0026279	0.0065904	0.006415
	0.076	0.003625	0.0021487	0.006849	0.003957
	0.109	0.01	0.003355	0.044	0.00946
	0.185	0.003710	0.000769	0.041	0.001737
Run 2	0.043	0.0346	0.0163	0.0561	0.05357
	0.076	0.03307	0.01345	0.05568	0.03582
	0.109	0.01347	0.005683	0.0454	0.0175
	0.185	0.00414	0.031419	0.031419	0.00314
Run 3	0.043	0.03149	0.016447	0.0044629	0.05091
	0.076	0.021	0.0224	0.0686	0.06
	0.109	0.012	0.0474	0.0555	0.0126
	0.185	0.0153	0.00678	0.0526	0.00206
Run 4	0.043	0.03793	0.01944	0.048708	0.0584
	0.076	0.031293	0.016273	0.049875	0.048337
	0.109	0.003176	0.00088	0.016128	0.002539
	0.185	0.009607	0.002491	0.049959	0.0001095

Table 3.9. Constants *a* and *b* in Rayleigh distribution function for acetylene

Run No.	Run 1	Run 2	Run 3	Run 4	Sunderland and Feath	Onischuk et al.
a	0.02789	0.02003	0.01869	0.0176	0.0048	0.01154
b	270.416	200	138.89	130.07	800	400

3.2.6.2. Soot Particles

Experiments of synthesis carbon blacks are carried out to develop to analysis the yields of soot particles. Gas is sampled through a tap at the height level of 0.45m on the combustion chamber and is analyzed. Operational conditions and measured gas concentrations are given in Table 3-10. The yields of soot particles obtained by element balance from the measured gas concentration in two ways.

(1) When all carbons are converted into to CO and CO_2, and hydrogen is converted to H_2O, gasification reactions dominate and the overall gasification, reactions is written as

$$0.717N_2 + AC_3H_8 + 0.191O_2 \rightarrow aC + bCO + cCO_2 + 4AH_2O + dO_2 + 0.717N_2 \tag{3-96}$$

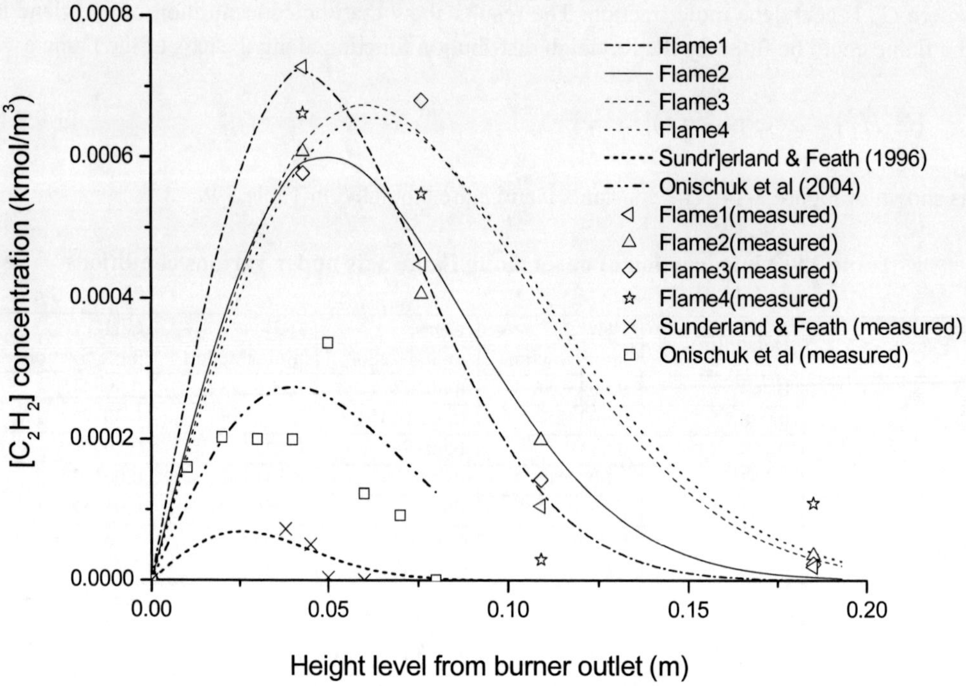

Figure 3.14. Acetylene profiles measured and fitted by the Rayleigh distribution function.

Table 3.10. Operational conditions and measured gas concentrations

C_3H_8 (NL/min)	C_3H_8 (mol/min)	V_{co_2}/V_t	V_{co}/V_t	V_{o_2}/V_t	Total O_2 (mol/min)	Total N_2 (mol/min)
3	0.0466	0.109	0.03	0.17	0.191	0.717
4	0.062	0.058	0.0123	0.114	0.191	0.717
5	0.0775	0.0615	0.077	0.0185	0.191	0.717
6	0.093	0.0397	0.0217	0.0162	0.191	0.717
6.5	0.10075	0.024	0.0185	0.0295	0.191	0.717

V_t is the total gas mole flow rate.

From element balance, it has

$$3A = a + b + c \tag{3-97}$$

$$0.191 = 0.5b + c + 2A + d \tag{3-98}$$

$$V_{CO} / V_{all} = \frac{b}{b + c + d + 0.717} \tag{3-99}$$

$$V_{CO_2} / V_{all} = \frac{c}{b + c + d + 0.717} \tag{3-100}$$

$$V_{O_2} / V_{all} = \frac{d}{b + c + d + 0.717} \tag{3-101}$$

where A is mole flow rate of propane.

(2) At higher C/O mole ratios, pyrolysis reactions occur, part of hydrogen is converted to C_2H_4, the overall reactions is written as

$$0.717N_2 + AC_3H_8 + 0.191O_2 \rightarrow aC + bCO + cCO_2 + eC_2H_4 + (4A - 2e)H_2O + dO_2 + 0.717N_2 \tag{3-102}$$

From element balance, it has

$$3A = a + b + c + 2e \tag{3-103}$$

$$0.191 = 0.5b + c + d + \frac{4A - 2e}{2} \tag{3-104}$$

$$V_{CO} / V_{all} = \frac{b}{b + c + d + e + 0.717} \tag{3-105}$$

$$V_{CO_2} / V_{all} = \frac{c}{b + c + d + e + 0.717} \tag{3-106}$$

$$V_{O_2} / V_{all} = \frac{d}{b + c + d + e + 0.717} \tag{3-107}$$

The calculated results are tabulated in Table 3-11. The calculated results show that when the ratio of flow rates of propane to O_2 is smaller than 0.406, all hydrogen is converted to H_2O and the reactions are in gasification regime. When the ratio of flow rates of propane to O_2 is more than 0.406, part of hydrogen is converted to C_2H_4 and the reactions are in the pyrolysis regime as shown in Figure 3-15.

Table 3.11. Results of calculation

C_3H_8(NL/min)	C_3H_8/air	C(mol/min)	CO(mol/min)	CO_2(mol/min)	C_2H_4(mol/min)
3	0.0512	0.035	0.023	0.82	
4	0.068	0.13	0.0097	0.0458	
5	0.085	0.1904	0.0234	0.0102	
6	0.102	0.1204	0.019	0.0348	0.05024
6.5	0.1108	0.126	0.017	0.022	0.0685

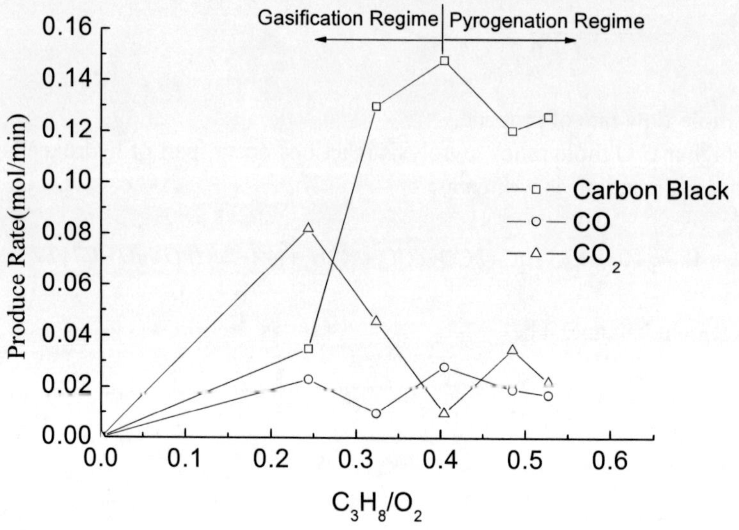

Figure 3.15. Effects of the ratio of flow rates of propane to O_2.

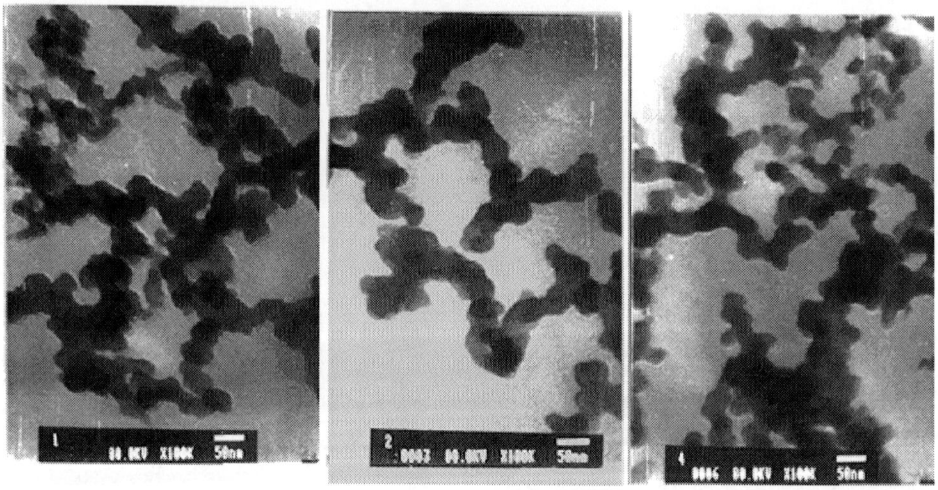

Figure 3.16. Typical TEM images of soot particles.

In the test runs listed in Table 3-6, a fine sieve of 600 mesh is placed at the top of the combustion chamber to avoid the soot particles emitted into the ambiance and to collect the soot particles for TEM measurements. A typical TEM image of soot particles is shown in Figure 3-16. The image demonstrates the clustering of primary soot particles to aggregates. The polydispersion even of primary particles, *e.g.*, the size of soot particles obtained from TEM pictures is also listed in Table 3-6 and is about 30nm for the four experimental runs.

The yields of soot particles obtained by element balance from the gas concentration measured through the tap at the position of 0.45m on the combustion chamber, which is higher than the flame length. CO_2 and CO are detected and measured, and soot yields are calculated through equations (3-96) to (3-101). The soot yield increased from 0.42g/min to 2.28g/min as the mole ratio of propane to air increased from about 0.05 to 0.085, and decreased from 2.28g/min to 1.448g/min as the mole ratio of propane to air increased further from 0.085 to 0.1 as can be seen from Table 3-6.

3.2.6.3. Surface Growth of Soot Particles

Soot volume fraction f_s is defined as the volume of soot particles per unit volume of gas:

$$f_s = \frac{\pi}{6} d_p^3 n_p \tag{3-108}$$

where d_p is the mean diameter of soot particles, n_p is number density of soot particles. For surface growth, the growth rate of soot particles, W_g, as the rate of increase of soot mass per unit soot surface area and time [11]

$$W_g = \rho_s v_g = (\rho / S) d(\rho_s f_s / \rho)/dt \tag{3-109}$$

where v_g is the soot growth velocity. The soot surface area per unit volume, S, in equation (3-109) can be found as

$$S = \pi d_p^2 n_p = 6 f_s / d_p \tag{3-110}$$

The gas density is found from present concentration and temperature measurements, assuming an ideal gas mixture of the major gas species and neglecting the volume of soot (which is only present at ppm levels). The soot density is taken to be ρ_s=1800kg/m^3.

As the growth of soot particles is associated with acetylene concentration, the rate of surface growth of soot particles, W_g is proposed as [11]

$$W_g = k_g(T)[C_2H_2]^n \tag{3-111}$$

experimental results have showed that the coefficient k_g has little dependence on temperature in the temperature range 1250-2100K [11] and the reaction order is about 0.9 and 1.65 for corrected and uncorrected with soot oxidation reaction, respectively [11]. In this work, a first order reaction is proposed for propane/air diffusion flames

$$W_g = \alpha [C_2H_2] \tag{3-112}$$

for different propane/air ratios.

3.2.6.4. Nucleation Rate of Soot Particles

The nucleation rate of soot particles, W_n, is defined as the rate of increase of the number of soot particles per unit volume and time, and is equal to [1]

$$W_n = \rho d(n_p / \rho)/dt \tag{3-113}$$

and is a first-order nucleation [11]

$$W_n = 10.6[C_2H_2]\exp(-16100/T) \tag{3-114}$$

3.2.6.5. Growth of Soot Particles

Neglecting the change of gas density in the flame, equations (3-113) and (3-114) give that

$$\frac{dn_p}{dt} = 10.6[C_2H_2]\exp(-16100/T) \tag{3-115}$$

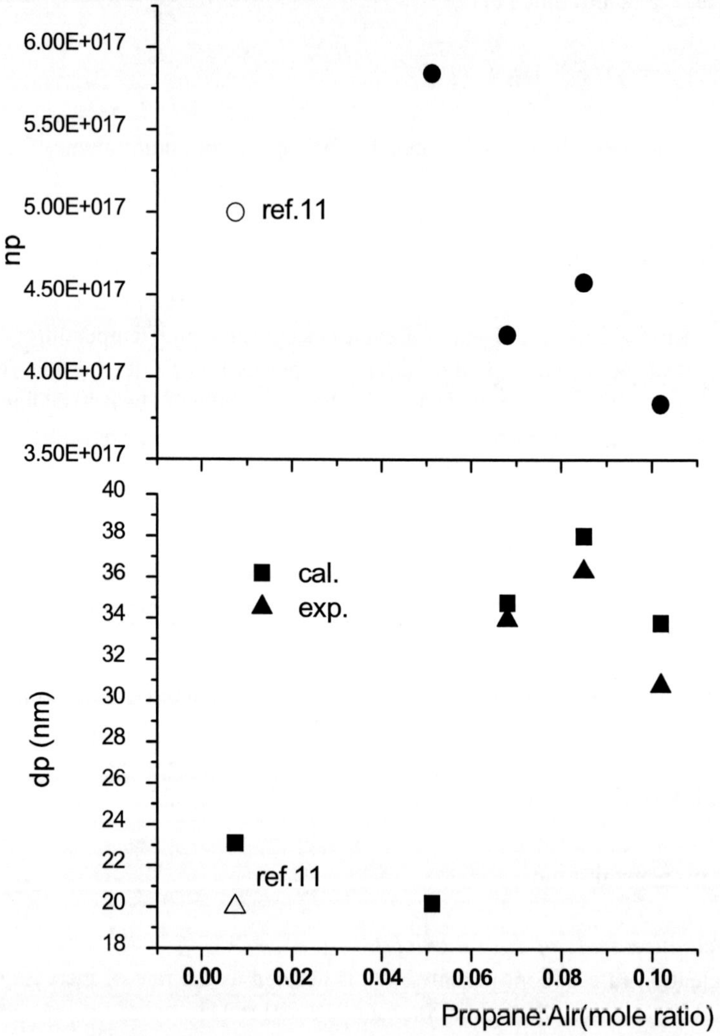

Figure 3.17. Values of d_p and n_p under different operate conditions ▲ soot size determined by TEM ■ soot size from calculation ● number density of soot particles.

Combining equations (3-108) to (3-110) and (3-112), one can obtain that

$$W_g = \frac{300}{n_p d_p^2}\left(d_p^3 \frac{dn_p}{dt} + 3n_p d_p^2 \frac{dd_p}{dt} \right) = \alpha[C_2H_2] \qquad (3\text{-}116)$$

the particle size d_p and number density n_p can then be calculated from equations (3-115) and (3-116) at given operational conditions and the rate coefficient of surface growth α. The calculated results of particle size d_p and number density n_p are shown in Figure 3-17 for the experimental runs, and the runs from literature [11]. Comparisons are made on particle size between calculated and measured. The particle size is measured by TEM method by averaging the particles counted from TEM photographs.

In the calculation for particle size and number density, the rate coefficient α is determined from that the calculated yields of soot particles are equal to the measured values. Since the yield of soot particles $\dot{m}$ (kg/s) is equal to

$$\dot{m} = \frac{\pi}{6}\rho_s d_p^3 n_p \dot{V} \qquad (3\text{-}117)$$

where $\dot{V}$ (m^3/s) is the total gas volume flow rate, the yields of soot particles can then be obtained from the calculated results of particle size and number density at a given value of the rate coefficient α. The rate coefficient α is such taken as the calculated yield of soot particles is equal to the measured value. The rate coefficient of surface growth of soot particles α such determined are shown in Figure 3-18. The figure shows that the rate coefficient α increases with increasing the mole ratio of propane to air and is almost level off at about 0.35 when the mole ratio of propane to air is more than about 0.07.

3.3. Simulation of Particle Growth

Since the importance of size distribution and difficulties of its mathematical descriptions, simulation of size distribution has received extensive studies. The development of the size distribution by coagulation can be described by the coagulation equation proposed by Muller [16], which is a nonlinear integral differential equation. A number of methods e.g., sectional technique, self-preserving model, has been developed to solve the coagulation equation numerically, which has been reviewed by Fission [17]. The main disadvantage of Muller's method is that it is difficult to integrate it with the mass, momentum and energy equations of gas domain for process design, which is encountered with aerosol processes.

From the viewpoint of the balance of particle number density, the development of the size distribution by coagulation in aerosol processes could be treated by the principle of fluid dynamics. The coagulation process could then been simulated by implementation of a coagulation model in computational fluid dynamics (CFD) by Kruis model [18]. And dilution effects caused by non-uniform temperature or gas density have also been counted by Johannessen [19]. In the Kruis and Johanessen models, the change rate of particle number density due to their collisions is calculated from the collision kernel by Fuchs' interpolation equation [20].

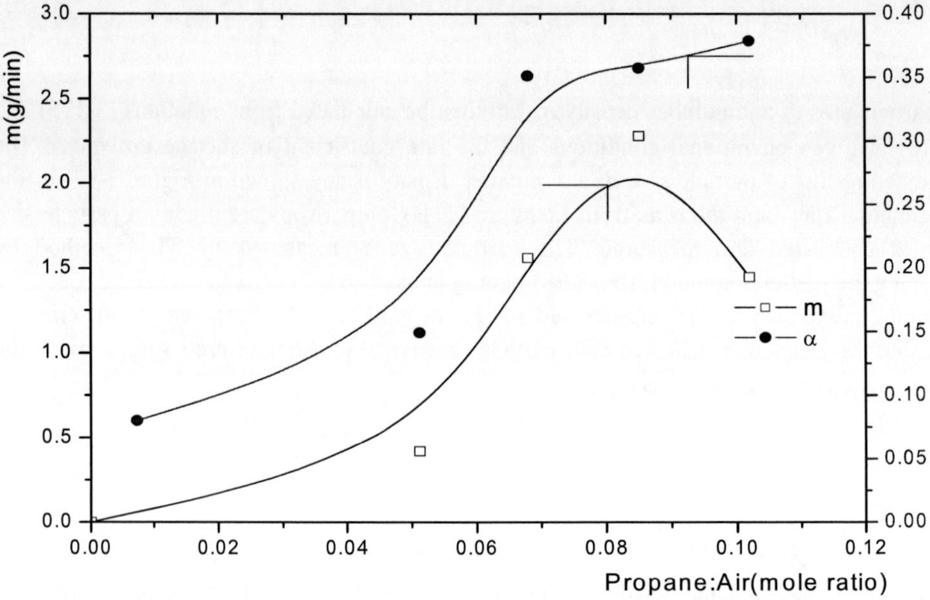

Figure 3.18. Soot yield and surface reaction coefficient *vs* mole ratio of propane to air.

3.3.1. Simulation Methods

The decrease rate of particle/aggregate number density due to particle coalescence is assumed to be

$$\frac{dN}{dt} = -\beta N^2 \tag{3-118}$$

If we assume that particle volume density is constant in coagulation, *i.e.*,

$$N\upsilon = \text{constant} \tag{3-119}$$

where υ is particle volume. Differentiating equation (3-119) and substituting equation (3-118) give

$$\frac{d\upsilon}{dt} = \beta N\upsilon \tag{3-120}$$

If we assume that particle surface area density is constant in coagulation, *i.e.*,

$$Na = \text{constant} \tag{3-121}$$

where a is particle surface area. Differentiating equation (3-121) and substituting equation (3-118) give

$$\frac{da}{dt} = \beta Na \tag{3-122}$$

If we assume that particle entity $Nd^{2.5}$ is constant in coagulation, *i.e.*,

$$Nd^{2.5} = \text{constant} \tag{3-123}$$

Differentiating equation (3-123) and substituting equation (3-118) give

$$\frac{dd}{dt} = \frac{1}{2.5}\beta Nd \tag{3-124}$$

The velocity components, temperature, pressure, etc, of the gas-phase domain are first calculated by CFD solver, Fluent. The calculations for particle entities such as number density N, volume density $N\upsilon$ or surface area density Na are implemented into the CFD codes as

$$\frac{\partial}{\partial x_k}\left(\rho_g u_k \phi - D\frac{\partial \phi}{\partial x_k}\right) = S \tag{3-125}$$

where ϕ and S stand for a particle entity and its source term, respectively, D is particle's diffusion coefficient. The particle entities ϕ and their source terms S are listed in Table 3-12.

Under the assumption that the reaction of precursor with O_2 happens instantaneously at the outlet of the burner, converting all the precursor molecules to monosized monomers of 1 nm, the particle number density at the outlet of burner, N_0, is equal to

$$N_0 = \frac{6}{\pi} * 10^{27} \kappa \frac{M_P}{M_{pr}} \frac{\dot{w}_{pr}}{\dot{V}\rho_P} \tag{3-126}$$

where κ is the stoichiometric ratio of a reaction, M_P and M_{pr} are the molecular weights of product and precursor, respectively, $\dot{w}_{pr}$ is the mass flow rate of precursor, and $\dot{V}$ the total gas volumetric flow rate.

3.3.2. Simulation by Kruis Model

Initially, the particle coalescence rate is much larger than the collision rate so spherical particles are formed. As the particle size increases, however, aggregates of spherical primary particles are formed as demonstrated in Figure 3-19, and assuming that all aggregates contain the same number of equally sized primary particles.

Table 3.12. Particle entities ϕ and their source terms S

particle entity	N	υ	a	$N\upsilon$	Na	$Nd^{2.5}$	$\rho_g N$
source term	$-\beta N^2$	$\beta N\upsilon$	βNa	0	0	0	$-\rho_g \beta N + \frac{N\mu_t}{m_{PM} Sc_t}\Delta m_{PM}$ *

* When dilution effects are accounted.

The surface area of an aggregate particle, *a,* increases by coagulation and decreases by sintering is

$$\frac{da}{dt} = -\frac{1}{N}\frac{dN}{dt}a - \frac{1}{\tau}(a - a_s) \tag{3-127}$$

where τ is the characteristic sintering time and a_s is the surface area of the completely fused aggregate

$$a_s = \left(\frac{\upsilon}{\upsilon_0}\right)a_0 \tag{3-128}$$

with a_0 and υ_0 denoting monomer surface area and volume, respectively. The volume of an aggregate, υ, is only affected by coagulation, e.g. equation (3-119) still holds.

The primary particle size, d_p, and the number of primary particles, n_p, in an aggregate are

$$d_p = \frac{6\upsilon}{a} \tag{3-129}$$

$$n_p = \frac{\upsilon}{\upsilon_p} \tag{3-130}$$

where υ_p is the volume of a primary particle

$$\upsilon_p = \frac{\pi}{6}d_p^3 \tag{3-131}$$

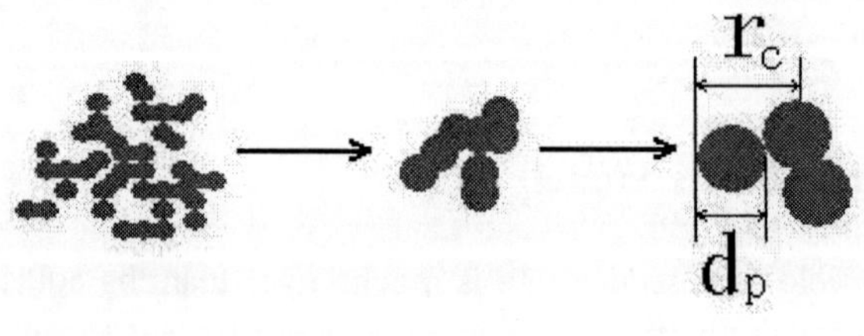

Figure 3.19. Illustration of aggregate generation phenomena.

The collision kernel β of the Fuchs interpolation in the free molecule and continuum regimes is given by

$$\beta = 8\pi D r_c \left[\frac{r_c}{2r_c + \sqrt{2}g} + \frac{\sqrt{2}D}{c r_c} \right]^{-1} \tag{3-132}$$

where r_c is the collision radius for irregular shaped aggregates, D is the particle diffusion coefficient, c is particle velocity, and g is transition parameter, which are given by

$$r_c = r_p \left(n_p\right)^{\frac{1}{D_f}} \tag{3-133}$$

$$D = \frac{k_b T}{6\pi\mu r_c} \left[\frac{5 + 4K_n + 6K_n^{\,2} + 18K_n^{\,3}}{5 - K_n + (8+\pi)K_n^{\,2}} \right] \tag{3-134}$$

$$c = \sqrt{\left(\frac{8k_b T}{\pi \rho_p v} \right)} \tag{3-135}$$

$$g = \left[\frac{1}{6r_c l} \right] \left[(2r_c + l)^3 - \left(4r_c^{\,2} + l^2\right)^{\frac{3}{2}} \right] - 2r_c \tag{3-136}$$

where

$$K_n = \frac{\lambda}{r_c} \tag{3-137}$$

$$l = \frac{8D}{\pi c} \tag{3-138}$$

and r_p is the radius of the primary particle (=$d_p/2$) and D_f is the mass fractal-like dimension.

If the number of primary particles in the aggregate reduces to one, the collision radius r_c becomes equal to the primary particle radius; for more than one primary particle per aggregate, the collision radius approaches the radius of gyration of the aggregate as illustrated in Figure 3-19.

When the dilution effects are accounted due to that gas density changes with temperature change [19], the decrease rate of particle/aggregate number density due to particle coalescence is then assumed to be

$$\frac{dN}{dt} = -\beta N^2 + \frac{N}{f_D}\frac{df_D}{dt} \tag{3-139}$$

where f_D is a dilution factor

$$f_D = \frac{m_{PM}}{m_{PM,0}} \tag{3-140}$$

where m_{PM} is the mass fraction of particles in the gas domain where the particles is taken as pseudo gas component and subscript *0* denotes the conditions at the inlet.

The species transportation equation for the pseudo gas component of particles is then

$$\nabla(\rho_g \mathrm{vm}_{PM}) = \nabla(\frac{\mu_t}{Sc_t}\nabla \mathrm{m}_{PM}) \tag{3-141}$$

Multiplying gas density ρ_g in the both sides of equation (3-139) and combining with equations (3-140) and (3-131), the particle/aggregate number density balance equation becomes

$$\frac{d(\rho_g N)}{dt} = -\rho_g \beta N^2 + \frac{N\mu_t}{m_{PM} Sc_t}\Delta m_{PM} \tag{3-142}$$

3.3.3. Simulation by Integral Collision Kernel

The reaction of precursor with O_2 is assumed to occur instantaneously at the outlet of the burner, converting all the precursor molecules to product monomer molecules.

The subsequent formation and growth of particles/aggregates occur by coagulation caused by collisions between particles/aggregates with monomers or between particles/aggregates, leading to steadily growing particle size and size distribution.

It is assumed that a new spherical particle is formed once particles or aggregates are collide, which is consistent with TEM images as the particles are predominantly spherical as shown in Figure 3-20.

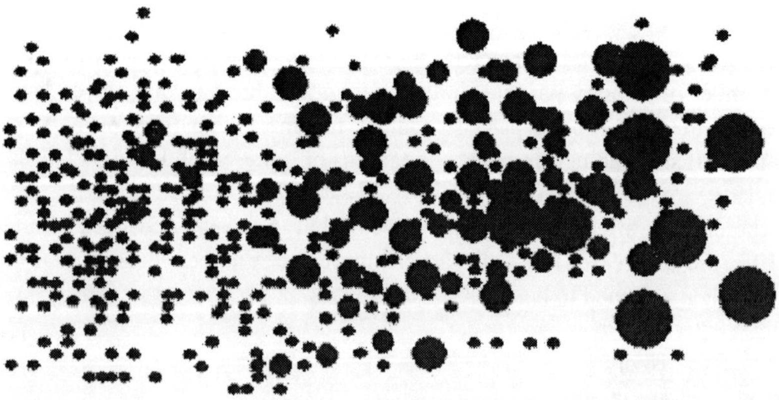

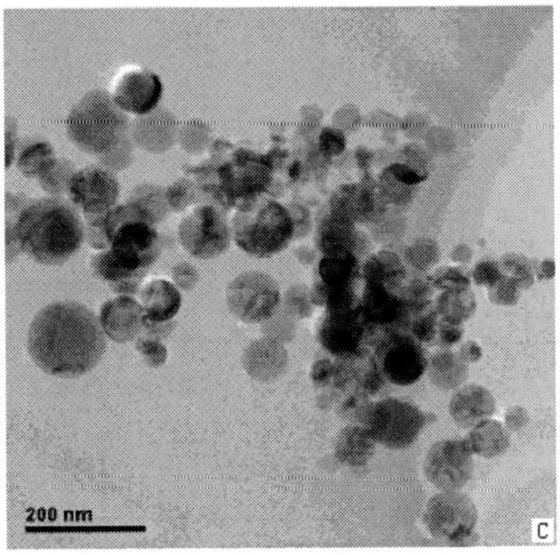

Figure 3.20. Illustration of particle growth phenomena.

In this case, the integral collision kernel is discussed in the section 2.2.4 of Chapter 2, the integral collision kernel in transition regime β_{TR} is given by equations (2-76) to (2-79), and repeated here as

$$\beta_{TR} = \sqrt{\frac{kT\lambda}{\rho_P}} h(Kn) \tag{3-143}$$

where *h(Kn)* is a fitting function and is obtained by the interpolation between the two limiting regimes as

$$h(Kn) = \frac{\gamma \sinh(5Kn)}{1 + \frac{\gamma \sinh(5Kn)}{p(Kn)}} \tag{3-144}$$

where sinh*(5Kn)* is the hyperbolic function, γ and *p(Kn)* are

$$\gamma = \sqrt{\frac{kT\rho_P}{\lambda\mu^2}} \tag{3-145}$$

and

$$\mathrm{p}(\mathrm{Kn}) = \frac{8.495}{\sqrt{\mathrm{Kn}}} + \frac{8.495}{\mathrm{Kn}^2} \tag{3-146}$$

respectively.

The particle's diffusion coefficient *D,* taken as the Stokes-Einstein diffusion coefficient, is given in equation (2-41) and is repeated here as

$$D_P = \frac{C_C}{3\pi\mu d} kT \tag{3-147}$$

and C_C is the Cunningham slip factor, which is given in equation (2-34) and is repeated here as

$$C_C = 1 + 2Kn\left(1.257 + 0.4e^{-0.55/Kn}\right) \tag{3-148}$$

3.4. Simulation Results

Size distributions of the sixteen experimental runs discussed in Chapter 1 are simulated.

The experimental conditions are given in Table 1-1 in Chapter 1. Comparisons on the size distribution of cumulative fraction undersize versus size between measured and simulated are shown in Figure 3-21 for the sixteen experimental runs.

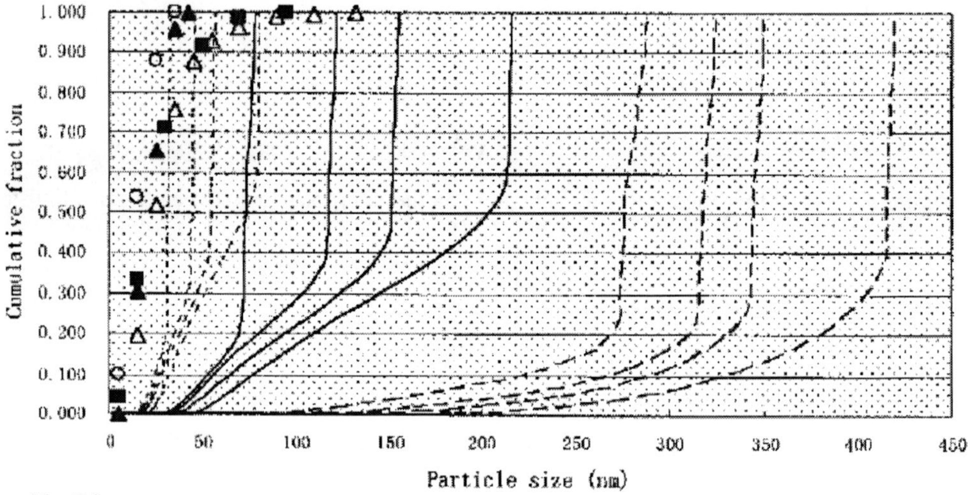

runs No. 5-8.

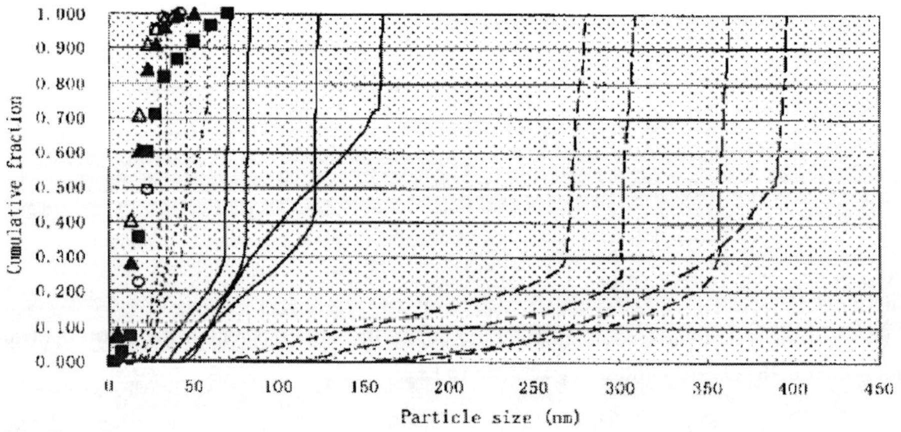

runs No. 1-4.

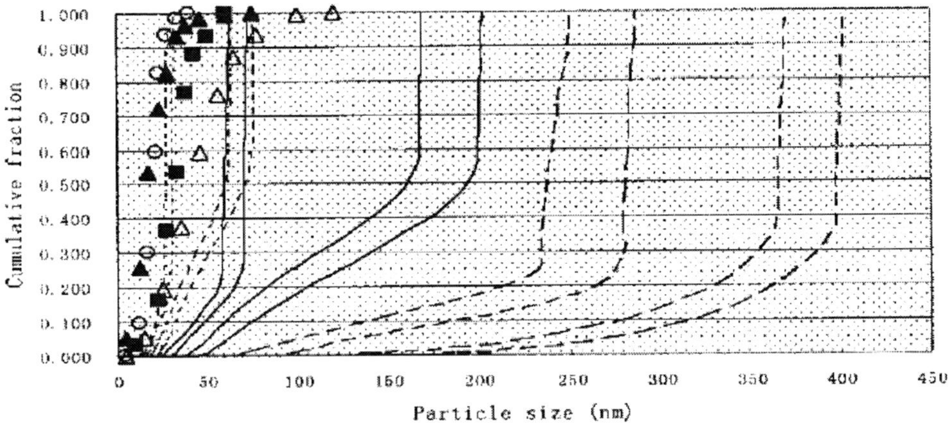

runs No. 9-12.

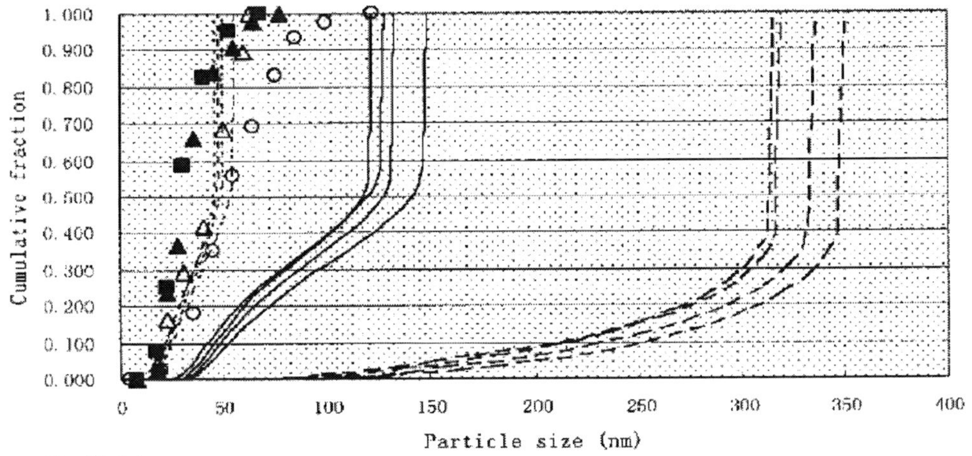

runs No. 13-16.

Figure 3.21. A comparison on the size distribution of cumulative fraction undersize versus size between measured and simulated by $N\upsilon$=constant, $Nd^{2.5}$=constant and Na=constant.

It could be seen that simulations with $N\upsilon$=constant give much better results than that with Na=constant and $Nd^{2.5}$=constant. Comparisons on median size and relative spread of size distributions between measured and simulated are presented in Table 3-13.

The table also shows that simulations with $N\upsilon$=constant give reasonable predictions on mean particle and size distributions for the sixteen experimental runs. The relative spread of size distributions is defined as

$$\delta = \frac{d_{84\%} - d_{16\%}}{2d_{50\%}} \tag{3-149}$$

Comparisons on the size distribution of number frequency between measured and predicted by integral collision kernel with $N\upsilon$=constant and by the Kruis model are shown in Figure 3-22 for the sixteen experimental runs. The figure shows that simulations by integral collision kernel with $N\upsilon$=constant give better predictions.

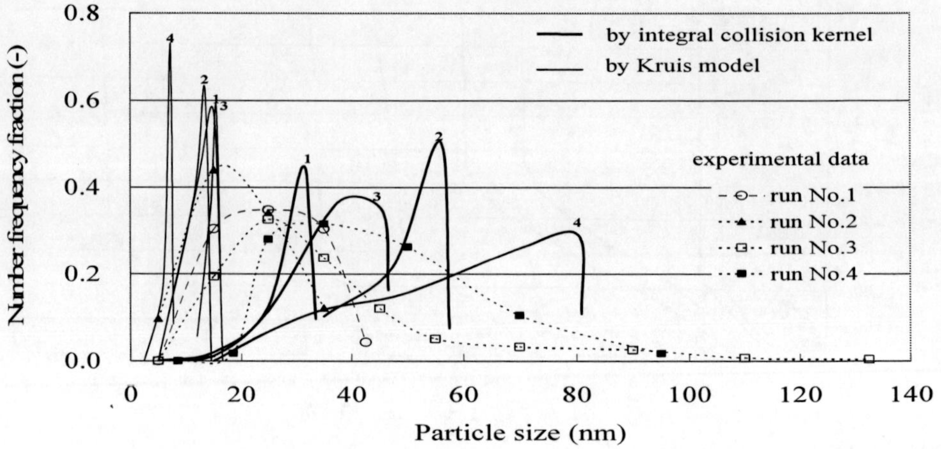
by integral collision kernel
by Kruis model
experimental data
run No.1
run No.2
run No.3
run No.4
Number frequency fraction (-)
Particle size (nm)

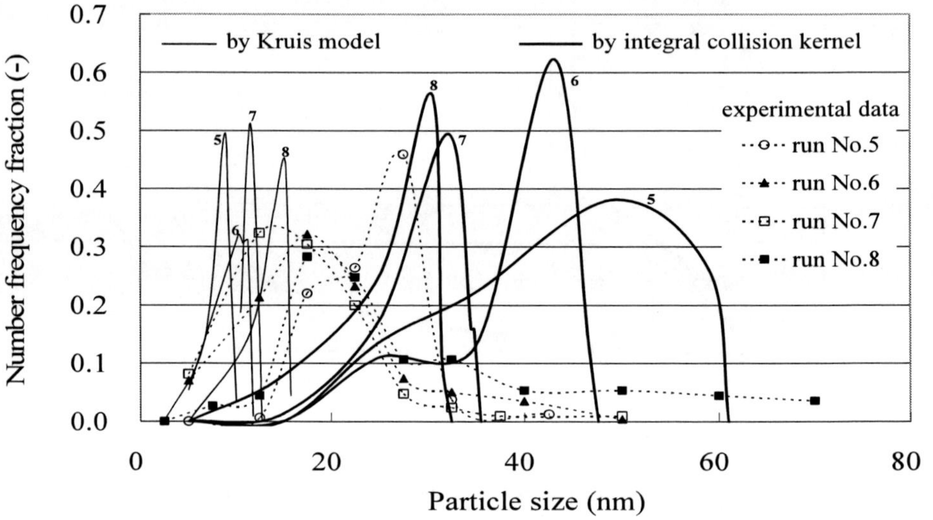
by Kruis model
by integral collision kernel
experimental data
run No.5
run No.6
run No.7
run No.8
Number frequency fraction (-)
Particle size (nm)

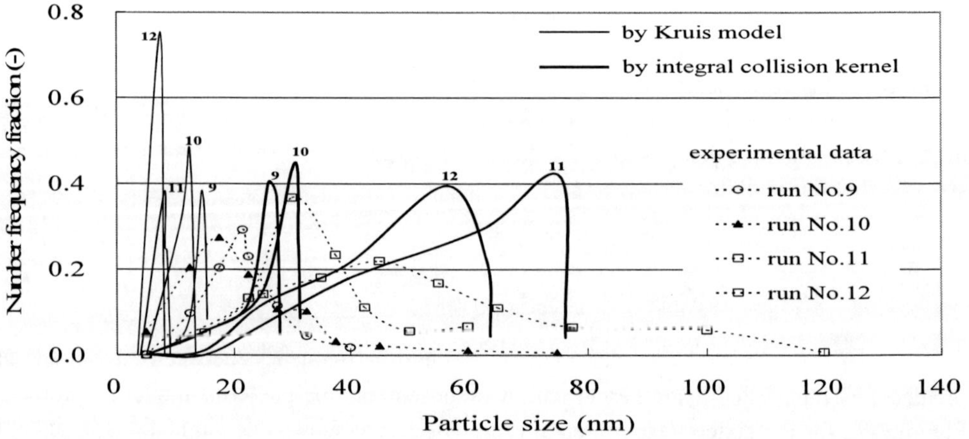
by Kruis model
by integral collision kernel
experimental data
run No.9
run No.10
run No.11
run No.12
Number frequency fraction (-)
Particle size (nm)

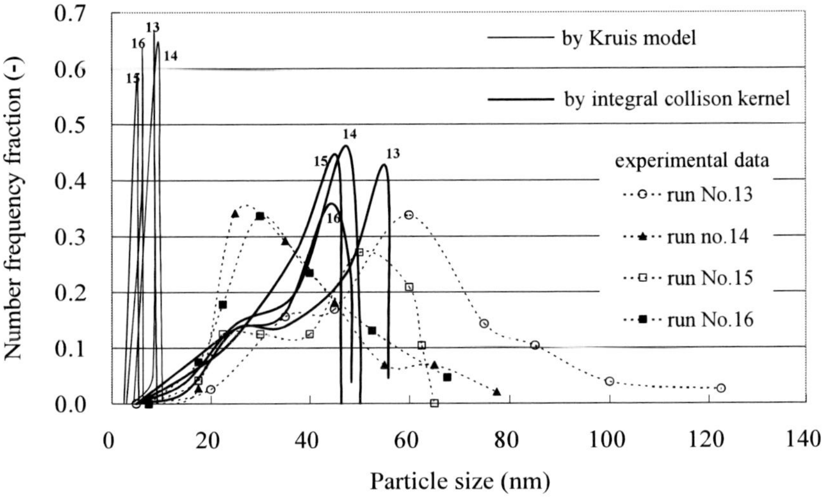

Figure 3.22. A comparison on the size distribution of number frequency between measured and simulated.

Table 3.13. Comparison on median size and relative spread of size distributions between measured and simulated

Run No.	Measured		$N\upsilon$=constant		Na=constant		$Nd^{2.5}$=constant	
	Mean size (nm)	Relative spread	Mean size (nm)	Relative spread	Mean size (nm)	Relative spread	Mean size (nm)	Rela-tive spread
1	24.3	0.500	31.7	0.068	276.1	0.041	73.3	0.090
2	16.6	0.586	55.9	0.190	345.0	0.043	151.6	0.251
3	32.3	0.612	45.0	0.150	318.0	0.039	117.0	0.218
4	42.9	0.500	74.0	0.294	415.0	0.066	204.1	0.294
5	23.6	0.227	47.0	0.300	387.0	0.124	120.0	0.392
6	23.5	0.453	45.9	0.133	358.0	0.049	121.0	0.202
7	16.6	0.500	34.1	0.095	302.3	0.061	81.2	0.135
8	24.8	0.537	30.1	0.106	271.4	0.114	68.9	0.139
9	26.0	0.155	26.8	0.070	237.0	0.137	59.0	0.074
10	16.5	0.559	30.8	0.074	281.0	0.082	70.8	0.099
11	61.9	0.476	71.6	0.272	398.0	0.065	195.0	0.279
12	56.7	0.313	61.0	0.254	366.0	0.070	162.0	0.287
13	56.5	0.385	53.0	0.255	347.0	0.081	140.1	0.296
14	48.9	0.593	49.0	0.230	333.3	0.094	127.5	0.294
15	37.7	0.394	46.4	0.218	317.1	0.093	119.0	0.277
16	39.3	0.333	47.2	0.244	313.5	0.096	120.5	0.311

References

[1] Fluent Inc., Fluent: *User's Guide* 6.0, 2002.

[2] Casimir, J. (1984). Chemical kinetic reaction mechanism for the combustion of propane, *Combust. Flame*, 55:213.

[3] Werborg, B.L., Howard, J.B., William G.C. (1973). Physical mechanisms in carbon formation in flames. *Proc. Combust. Inst.*, 14:929.

[4] Hepp, H., Siegmann, K. (1998). Mapping of soot particles in a weakly sooting diffusion flame by aerosol techniques, 115:275.

[5] di Stasio, S. (2001a). Electron microscopy evidence of aggregation under three different size scales for soot nano-particles in flame, *Carbon*, 39(1):109.

[6] Onischuk A.A, di Stasio, S., Karasev, V.V. et al. (2004). Evolution of structure and charge of soot aggregates during and after formation in a propane / air diffusion flame. *J. Aerosol Sci.,* 34:383.

[7] di Stasio, S., Mitchell, J.B.A., LeGarrec, J.L., Biennier, L., Wulff, M. (2006a). Synchrotron SAXS <in-situ> identification of three different size modes for soot nanoparticles in a diffusion flame, *Carbon*, 44(7):1267.

[8] di Stasio, S., Braun A. (2006b). Comparative NEXAFS study on soot obtained from an ethylene/air flame, a diesel engine, and graphite, *Energy and Fuels*, 20:187.

[9] Koylu,U.O., McEnally, C.S., Rosner, D. E., and Pfefferle, L.D. (1997). Simultaneous measurements of soot volume fraction and particle size/microstructure in flames using a thermophoretic sampling technique, *Combust. Flame*, 110: 494.

[10] Onischuk, A.A., di Stasio, S., Karasev, V.V., Baklanov, and Panfilov, V. N. (2001). Soot agglomerate convolution due to coulomb forces, *J. Aerosol Sci.*, 32(S1): 993.

[11] Sunderland, P.B. and Faeth, G.M. (1996). Soot formation in hydrocarbon / air laminar jet diffusion flames. *Combust. Flame*, 105: 132.

[12] Bockhorn, H., Fetting, F., Wannemacher, G., and Wenz, H.W. (1982). Optical studies of soot particle growth in hydrocarbon oxygen flames, *Nineteenth Symposium (international) on Combustion*, The Combustion Institute, Pittsburgh, pp. 1413.

[13] Harris, S.J., and Weiner, A.M. (1983). Soot formation in an ethylene diffusion flame, *Combust. Sci. Technol.* 31:155.

[14] Fusco, A., Knox-Kelecy, A.L., Foster, D.E. (1994). Application of a Phenomenological Soot Model to Diesel Engine Combustion, in The *Third International Symposium on Diagnostics and Modeling of Combustion in Internal Combustion Engine, COMODIA 94*, Yokohama, Japan, pp.571.

[15] Kronholm, D.F., Howard, J.B. (2000). Analysis of soot surface growth pathways using published plug-flow reactor data with new particle distribution measurements and published premixed flame data, *Proc. Combust. Inst.* 28: 2555.

[16] Muller H. (1928). Zur allgemeinen theorie der raschen koagula-tion, Kolloidbeihefte, 27, 223.

[17] Fissan H. J., Schoonman J. (1998). Vapor-phase synthese and processing of nanoparticle materials—A European Science Foundation (ESF) program. *J. Aerosol Sci.*, 29, 755.

[18] Kruis F.E., Kusters K.A., Pratsinis S.E., Scarlett B. (1993). A simple model for the evolution of the characteristics of aggregate particles undergoing coagulation and sintering, *Aerosol Sci. Technol.*, 19, 514.

[19] Johannessen T., Pratsinis S. E., Livbjerg H. (2000). Computational fluid-particle dynamics for the flame synthesis of alumina particles, Chem. Eng. Sci., 55, 177.

[20] Fuchs N.A. (1964). *Mechanics of Aerosols*, New York: Pergamon Press.

Chapter 4

PHOTOCATALYSIS FOR GAS PURIFICATIONS

NOMENCLATURE

b	adsorption coefficient ($=k_1/k_2$)
c	pollutant concentration, mgm^{-3}
D_A	diffusion coefficient of pollutant in air, m^2s^{-1}
L	effective length of reactor, m
k	1st order degradation rate constant, ms^{-1}
k_0	desorption rate, s^{-1}
k_1	adsorption rate constant, s^{-1}
N	number of layers
N_{end}	optimal number of layers
r	radial distance, m
$\dot{r}$	degradation rate, $mgm^{-2}s^{-1}$
R	inner diameter of reactor, m
Y	normalized concentration
z	axial distance, m

Greak Letters

α	parameter defined by equation (4-18)
β	$D_A\tau/R^2$
δ	effective film thickness, nm
δ_{max}	maximum effective film thickness, nm
γ	adsorption rate, s^{-1}
γ_0	desorption rate, s^{-1}
η	normalized axial distance z/L
ς	degradation degree, %
τ	residence time inside the reactor, s
θ	coverage of particles on the surface of quartz plate
ξ	normalized radial distance r/R

$\Delta\xi$	normalized radial step length

Subscripts

in	inlet
out	outlet

4.1. INTRODUCTION

Advanced oxidation processes (AOP) including ultraviolet (UV) radiation, ozone, hydrogen peroxide and/or catalyst, or their combinations are particularly useful for cleaning biologically toxic or non-degradable materials such as aromatics, pesticides, petroleum constituents, and volatile organic compounds in waste water and contaminated air [1]. Table 4-1 gives the oxidation potentials of oxidants of conventional and advanced oxidation technologies. AOP are designed to produce hydroxyl radicals, converting contaminant materials, to a large extent, into stable inorganic compounds such as water, carbon dioxide and salts [1].

Table 4.1. Oxidation potential of oxidants in conventional and advanced technology

Conventional oxidation technology		Advanced oxidation technology	
oxidants	Oxidation potential	oxidants	Oxidation potential
$Cl_2(ClO_2)$	1.36 V	·OH	2.8 V
$KMnO_4$	1.69 V	Methods to produce ·OH	plasma arc discharge photocatlytic
H_2O_2	1.77 V		
O_3	2.07 V		
F_2	2.87 V		

The most widely applied AOP have been H_2O_2/UV, O_3/UV, H_2O_2/O_3, H_2O_2/O_3/UV, Fenton, and photocatlytic oxidation (PCO) processes [1]. Semiconductor photocatalytic activities are based on the adsorption of an ultra band gap photon that promotes an electron from the valence band to the conduction band. The resulting electron-hole pair, after migration to the semiconductor surface, is then available for chemical redox reactions.

Semiconductor PCO processes have the advantages of photocatalytic oxidation at ambient temperature and pressure under sunlight or low-cost fluorescent light. Moreover, photocatalysts are generally nontoxic, inexpensive, and chemically and physically stable; final oxidation products are usually innocuous; and no additives are required [2-4]. Owing to its non-toxicity, high reactivity at ambient temperature, low cost, long-term stability and convenient band-gap energy, TiO_2-based photocatalyst is believed to be one of the most suitable photocatalysts for environment remediation [2,5-6].

Figure 4-1 illustrates a typical TiO_2 photocatalytic process of volatile organic compounds (VOCs). An electron in an electron-filled valence band (VB) is excited by photoirradiation to a vacant conduction band (CB), leaving a positive hole in the VB. These electrons and positive holes drive reduction and oxidation, respectively, of compounds adsorbed on the surface of a photocatalyst.

The activation equation can be written as

$$TiO_2+hv \rightarrow h^+ + e^- \qquad (4\text{-}1)$$

In this reaction, h^+ and e^- are powerful oxidizing and reducing agents, respectively. The oxidation and reduction reactions can be expressed as

$$\text{Oxidation reaction: } h^+ + OH^- \rightarrow OH\cdot \qquad (4\text{-}2)$$

$$\text{Reduction reaction : } e^- + O_{2ads} \rightarrow O_{2ads} \qquad (4\text{-}3)$$

When pollutant compounds are chemically transformed by a PCO device, it is the hydroxyl radical (OH·), derived from the oxidation of adsorbed water or adsorbed OH^-, that is the dominant strong oxidant.

Air purification studies of PCO process commonly use nanosemiconductor catalysts and ultraviolet (UV) light by using tubular photoreactors. A typical tubular photoreactor is illustrated in Figure 4-2.

The photoreactor is made up of a quartz tube, a UV lamps and a light shield. Thin film of nanophotocatalyst is coated on the inner wall of the quartz tube. A paper shade is attached to the outer wall of a quartz tube to diminish the loss of light radiation. Relative humidity of the incoming air is controlled by dividing the dry air into two routes, one passing through the deionizer water, and the other passing through a volatile organic compound (VOC) generator, and mixing the two effluents before entering the reactor. VOC loaded air stream is generated by the air passing through a VOC generator. A tube with an opening on the tube wall is installed into the VOC generator. The VOC is filled into the tube just below the opening, and the VOC concentration is controlled by changing the flow rate of the air passing through the VOC generator.

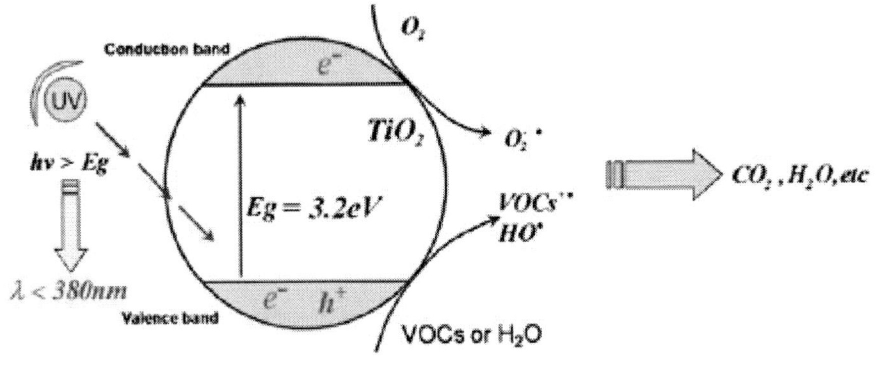

Figure 4.1. Illustration of TiO_2 photocatalytic process of VOCs.

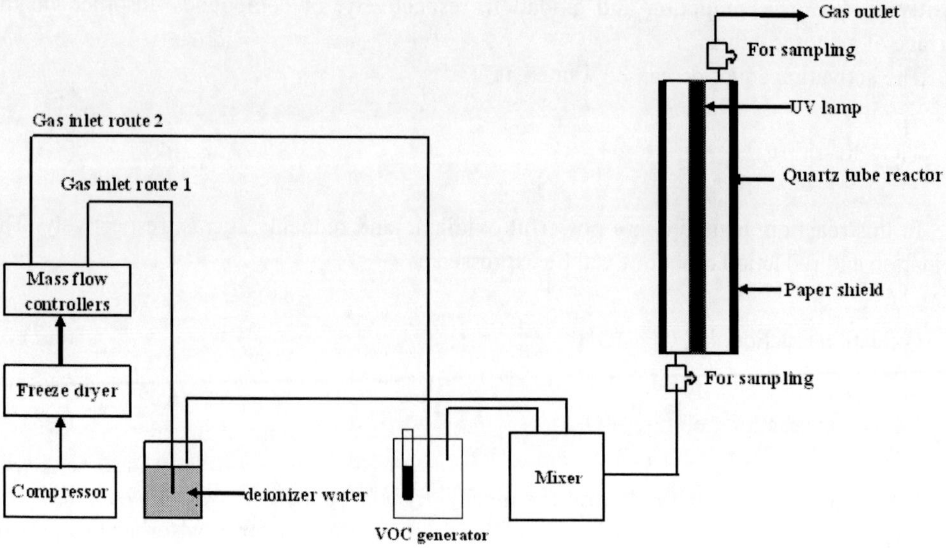

Figure 4.2. Illustration of tubular photocatalytic reactor for VOCs.

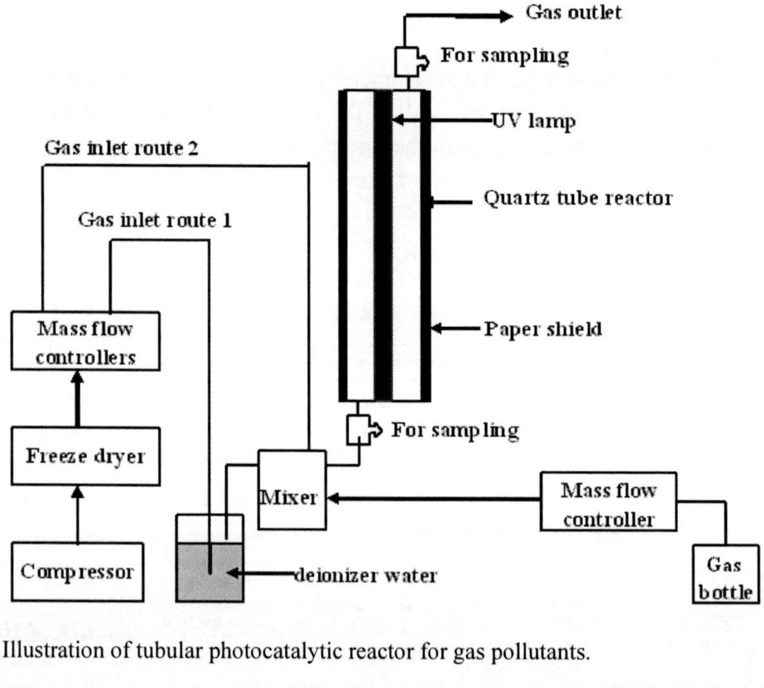

Figure 4.3. Illustration of tubular photocatalytic reactor for gas pollutants.

For gas pollutants, bottle gas is used and controlled by a mass flow controller instead of the VOC generator as shown in Figure 4-3.

The air stream is sampled every 30 minutes at both the inlet and outlet by using Tenax tubes at a flow rate of 0.5 L/min for analysis. Degradation degree ς can be then obtained by

$$\varsigma = \frac{c_{in} - c_{out}}{c_{in}} \qquad (4\text{-}4)$$

where c_{in} and c_{out} are pollutant concentration at inlet and outlet, respectively.

4.2. Preparation of Nanoparticles' Thin Film by Sendimentation

4.2.1. Effective Thickness of Thin Film

Thin film of nanoparticles is coated on a quartz plate by sedimentation. By doing this, a particle aqueous solution is prepared and vibrated by an ultrasonic vibrator for about 30 minutes to achieve the nanoparticles uniformly dispersed in the solution. A clean quartz plate (43mm×14mm×2mm) is then inserted into the particle aqueous solution. After leaving the quartz plate for a certain time, the quartz plate is taken out of the particle aqueous solution. One layer of the nanoparticles' thin film is then prepared by leaving the quartz plate in air for dry. More layers of the thin film coated on the quartz plate could be made by the operation repeatedly.

Figure 4-4 is a typical experimental result of adsorptive capacity of TiO_2 thin film on 365nm UV light versus number of TiO_2 thin film layers. In the figure, the concentration of TiO_2 aqueous solution is 2.5mmol/L, and sedimentation time is 3 hours. It can be seen that the adsorptive capacity increases with increasing the number of TiO_2 thin film layers and levels off after 9 layers of the TiO_2 thin film.

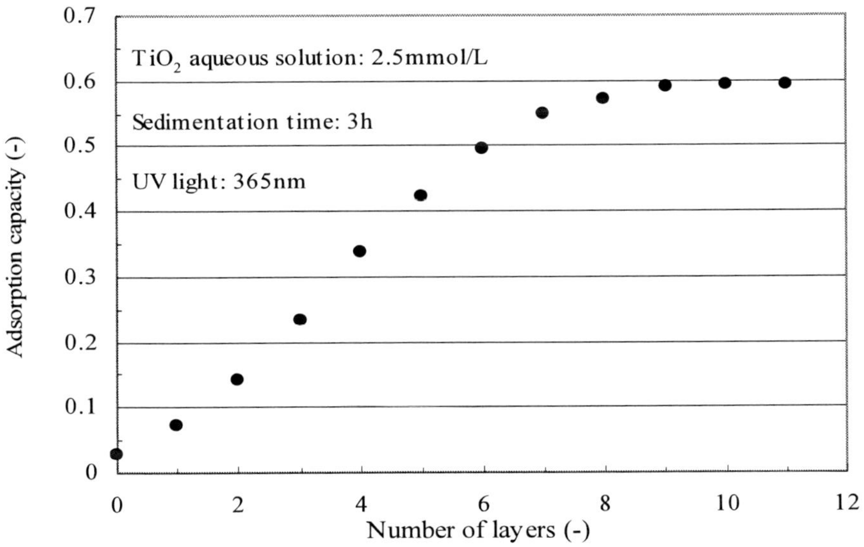

Figure 4.4. Adsorptive capacity on 365nm UV light versus number of TiO_2 thin film layers.

From the measured adsorptive capacity and TiO_2 mole adsorption coefficient by a UV spectrophotometer at a wavelength of about 365nm (its mole adsorption and percentage adsorption coefficients are about 1003.8L/cm/mol, and 125.6cm^{-1}), effective film thickness could be obtained. The effective film thickness is the thickness of a TiO_2 film, which has the same light transmission as the coated film of TiO_2 nanoparticles. The effective film thickness versus the number of TiO_2 thin film layers is shown in Figure 4-5. The figure shows that the effective film thickness increases up to about 275nm with increasing number of layers up to 9 layers, and then almost levels off as the number of layers further increases. Figure 4-6 is an AFM picture of 3 layers of TiO_2 thin film, it could be seen that there is still quartz surface not covered by the TiO_2 thin film.

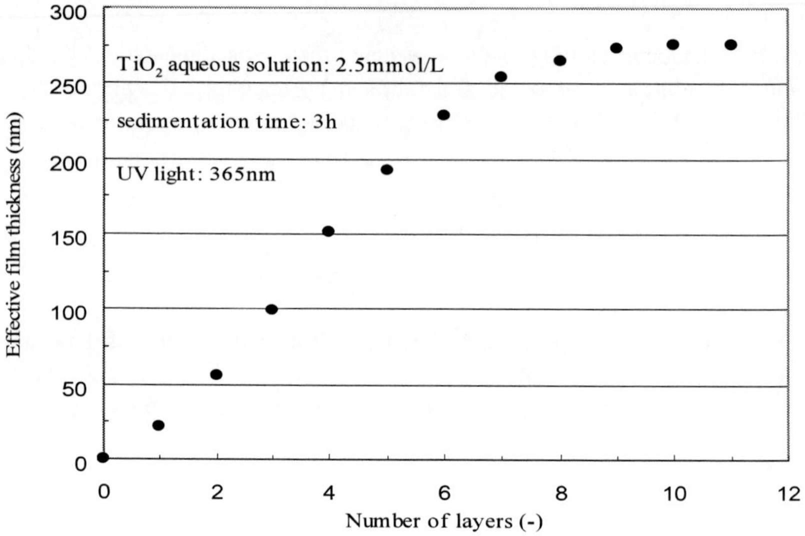

Figure 4.5. Effective film thickness versus number of TiO_2 thin film layers.

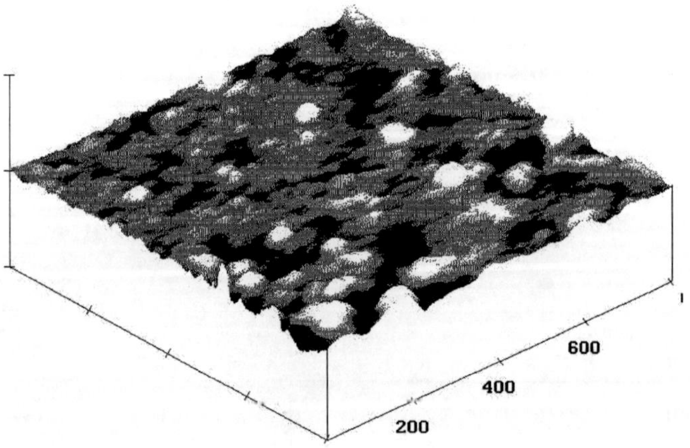

Figure 4.6. AFM picture of 3 layers of TiO_2 thin film (2.5mmol/L of TiO_2 aqueous solution 3 hours of sedimentation time).

4.2.2. Effects of Sedimentation Time

Effects of sedimentation time on effective film thickness are shown in Figure 4-7 at a TiO_2 aqueous solution of 0.125mmol/L. The film growths slowly at a short sedimentation time and growths fast at long sedimentation times as shown in Figure 4-7. When sedimentation time is over 1 hour, the film thickness reaches a maximum value of about 110nm. This may imply that adsorption and desorption reach equilibrium when sedimentation time is over 1 hour.

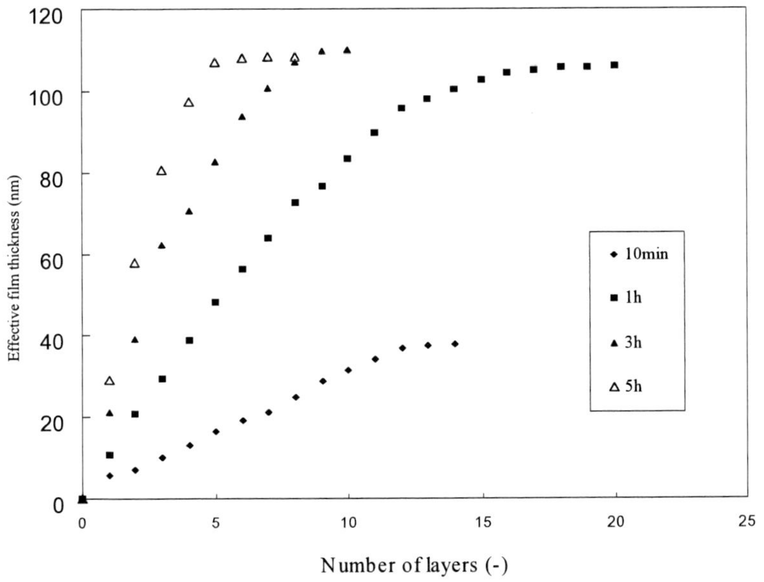

Figure 4.7. Effects of sedimentation time on effective film thickness.

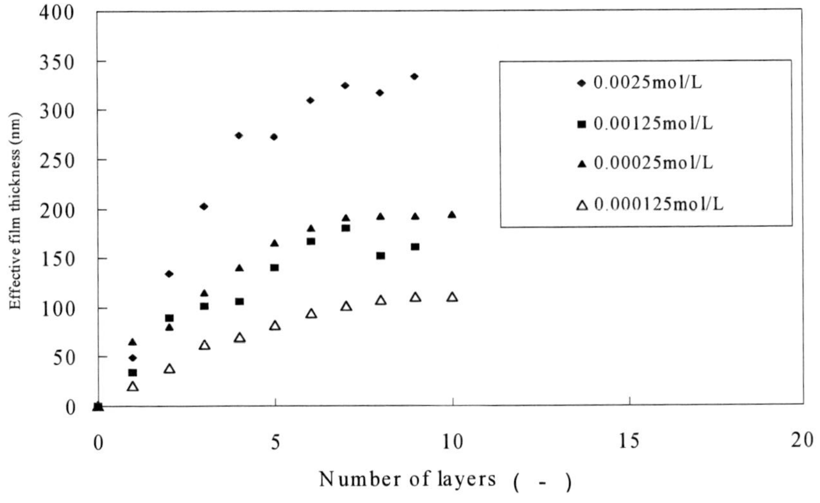

Figure 4.8. Effects of concentration of TiO_2 aqueous solution on effective film thickness (3 hours of sedimentation).

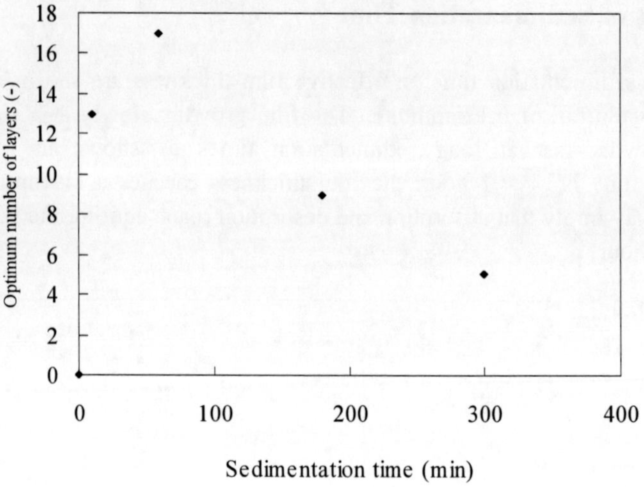

Figure 4.9. Optimal number of layers versus sedimentation time.

4.2.3. Effects of Concentration of TiO_2 Aqueous Solution

Effects of the concentration of TiO_2 aqueous solution on effective film thickness are shown in Figure 4-8 at the sedimentation time of 3 hours. The figure shows that as the concentration of TiO_2 aqueous solution increases, the film growths fast and the maximum film thickness increases.

The experimental results in Figures 4-7 and 4-8 yield that there exists a maximum film thickness and film thickness does not increase as number of layers further increases after the maximum film thickness has been reached. The number of layers at which the maximum film thickness approached is called optimal number of layers. It could be seen from figures 4-7 and 4-8 that TiO_2 concentration has little effects on the optimal number of layers, but sedimentation time has significant effects on the optimal number of layers. The effects of sedimentation time on the optimal number of layers is plotted in Figure 4-9, it shows that the optimal number of layers increases up to 17 when the sedimentation time increases up to about 1h, and then decreases as the sedimentation time further increases.

4.2.4. Analysis of Sedimentation Process by Adsorption Theories

The experimental results in Figure 4-7 show that the effective film thickness versus number of layers appears the blend of B.E.T and Langmuir adsorption isothermals at the sedimentation time of about 10min, and appears Langmuir adsorption isothermal at longer sedimentation times.

4.2.4.1. Sedimentation Process by Langmuir Adsorption Isothermal

Under the assumptions of that adsorption sites are uniformly distributed on the surface of quartz plate and each adsorption site adsorbs one particle, adsorption rate γ is then proportional to the concentration of adsorption sites on the surface of quartz plate and to the number of layers. If θ stands the coverage of particles on the surface of quartz plate, e.g.

$$\theta = \frac{\text{film thickness}}{\text{max ium film thickness at equilibriu m}} \quad (4\text{-}5)$$

then the concentration of adsorption sites on the surface of quartz plate will be proportional to (1-θ), thus

$$\gamma = k_1(1-\theta)N \quad (4\text{-}6)$$

where N is the number of layers.

Under the assumption of that adsorption sites are uniformly distributed on the surface of quartz plate, each adsorption site adsorbs one particle and there are no interactions between adsorbed particles, desorption rate γ_0 will be proportional to

$$\gamma_0 = k_2\theta \quad (4\text{-}7)$$

At equilibrium it has

$$\theta = k_1N/(k_2 + k_1N) = bN/(1+bN) \quad (4\text{-}8)$$

where $b=k_1/k_2$ is adsorption coefficient and N is the number of layers.

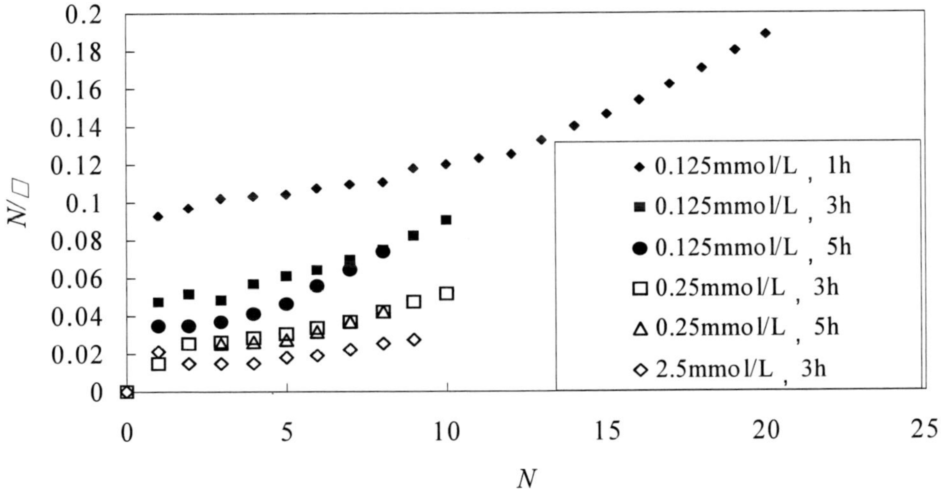

Figure 4.10. N/δ versus N for sedimentation process by Langmuir adsorption isothermal.

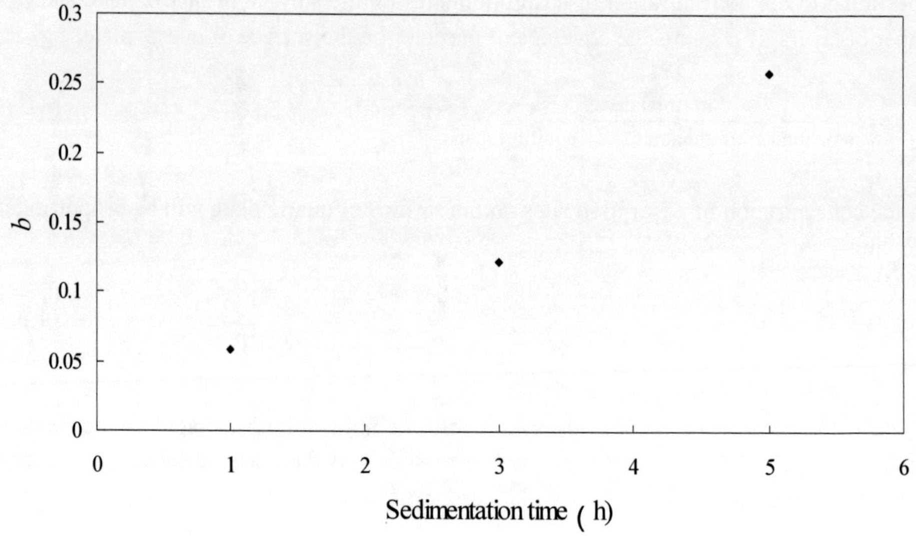

Figure 4.11. Effects of sedimentation time on parameter b (0.125mmol/L of TiO_2 aqueous solution).

4.2.4.2. Sedimentation Process by B.E.T Adsorption Isothermal

Similar to molecular adsorption process, by B.E.T adsorption isothermal one can obtained that

$$\delta = (\delta_{max} \cdot c \cdot N)/\{(N_{end} - N) \cdot [1 + (c-1)(N/N_{end})]\} \tag{4-9}$$

where N_{end} is the optimal number of layers, δ is effective film thickness, δ_{max} is the maximum effective film thickness. If $c>1$, equation (4-9) yields that the adsorption of the first layer dominate the sedimentation process, and the larger c is , the stronger effects the adsorption of the first layer has.

4.2.5. Parameters of Sedimentation Process

4.2.5.1. Parameters of Sedimentation Process by Langmuir Adsorption Isothermal

Since $\delta / \delta_{max} = \theta$, equation (4-8) can be rewritten as

$$N/\delta = 1/(\delta_{max} \cdot b) + N/\delta_{max} \tag{4-10}$$

Thus, plots of experimental data by N/δ and N as horizontal and vertical axes, respectively, should be linear lines as shown in Figure 4-10. By linear regression, parameters δ_{max} and b could be then obtained and are tabulated in Table 4-2. It can be seen from Table 4-2 that such obtained maximum film thickness is slightly higher than that measured. Parameter b increases as sedimentation time increases as shown in Figure 4-11.

Table 4.2. Parameters of Langmuir type sedimentation

Concentration (mol/L)	Sedimentation time	δ_{max} (nm)	δ_{max} (nm) measured	b
0.000125	1h	212.8	105.9	0.059
0.000125	3h	212.8	109.8	0.122
0.000125	5h	172.4	108.3	0.258
0.00025	3h	277.8	190.0	0.257
0.00025	5h	285.7	190.9	0.285
0.00125	3h	131.6	150.5	0.031
0.0025	3h	526.3	275.9	0.207

Table 4.3. Parameters for Langmuir -B.E.T type sedimentation

Particle concentration (mol/L)	Sedimentation time	Parameters for B.E.T type sedimentation			Parameters for Langmuir type sedimentation		
		δ_{max}(nm)	C	Num-ber of layers	δ_{max} (nm)	b	Num-ber of layers
0.000125	10min	9.93	15.030	1-9	131.6	0.031	9-13
0.00025	10min	105.5	0.200	1-6	256.4	0.028	6-14
0.00025	1h	52.9	189	5-9	769.2	0.020	9-17

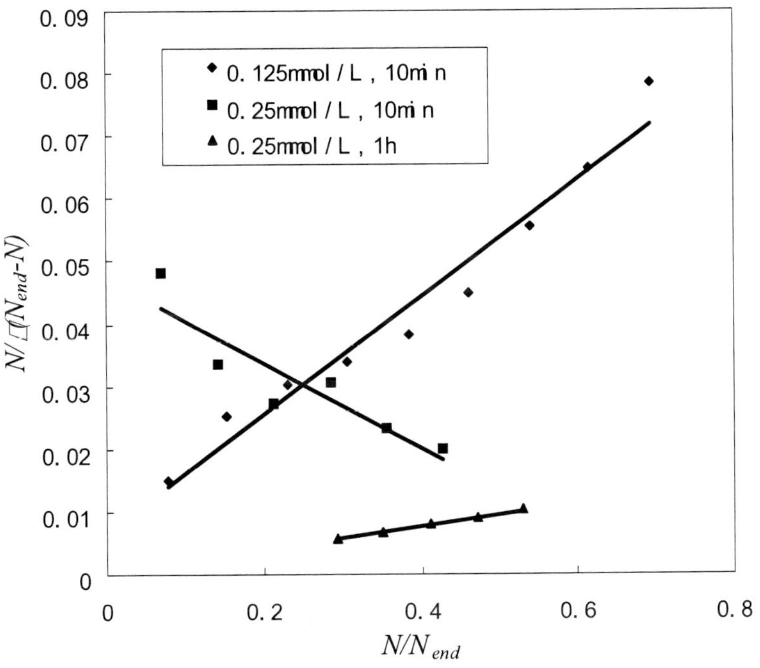

Figure 4.12. Parameters of sedimentation process by B.E.T adsorption isothermal.

4.2.5.2. Parameters of Sedimentation Process by Langmuir-B.E.T Adsorption Isothermals

The experimental results in Figure 4-7 show that the effective film thickness versus number of layers appears the blend of B.E.T. and Langmuir adsorption isothermals at short sedimentation times, and appears Langmuir adsorption isothermal at long sedimentation times. By rearranging equation (4-9) one can obtained that

$$N/[\delta \cdot (N_{end} - N)] = 1/(\delta_{max} \cdot c) + [(c-1)/(N_{end} \cdot c)] \cdot (N/N_{end}) \tag{4-11}$$

Thus, plots of experimental data by $N/[\delta \cdot (N_{end} - N)]$ and N/N_{end} as horizontal and vertical axes, respectively, should be linear lines for B.E.T. adsorption isothermal as shown in Figure 4-12. From equation (4-10), plots of experimental data by N/δ and N as horizontal and vertical axes, respectively, should be linear lines for Langmuir adsorption isothermal as shown in Figure 4-13.

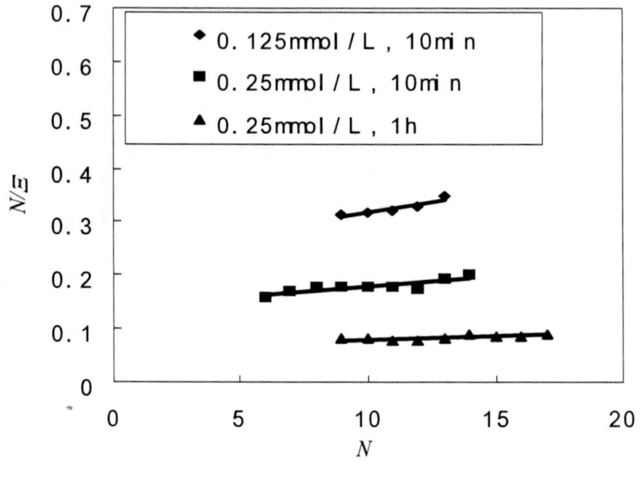

Figure 4.13. Parameters of sedimentation process by Langmuir adsorption isothermal.

Parameters δ_{max} and c could be then obtained by linear regression and are tabulated in Table 4-3. From Table 4-3 it can be seen that effects of the adsorption of first layer become stronger as the concentration of TiO_2 aqueous solution and sedimentation time increase.

4.3. PHOTODEGRADATION OF VOCS BY CARBON DOPED TiO_2 NANOPARTICLES

Several attempts have been made to increase the photodegradation efficiency of TiO_2 by noble metal deposition, ion doping, surface sensitizing with dyes and metals, adsorbent loading and etc [7-9]. Various adsorbents have been loaded to increase the photocatalytic efficiency of TiO_2, such as fiber glass, silica, alumina, zeolites, zinc oxide, active carbon, carbon black and carbon nanotubes [10-14]. These adsorbents make a high concentration environment of pollutants around TiO_2, hence, they increase the collision probability between pollutants and TiO_2. Additionally, it is reported that the adsorbent carbon nanotubes can also reduce the recombination of electron/hole exited by TiO_2 under Ultra Violet (UV) light, and the 1st order rate constant of the degradation of azo dye by Degusa TiO_2 photocatalyst P25 mixed with carbon nanotubes is about twice of that by P25 alone [14].

A tubular photoreactor used in the experiments is slightly different from that shown in Figure 4-2. The photoreactor is made up of a quartz tube, UV lamps and a light shield. The quartz tube has an inner diameter of 7.6mm, an outer diameter of 10mm, and has an effective length of 550mm. The inner wall of the quartz tube is coated with thin films of TiO_2 nanoparticles by sedimentation. Two UV lamps are used and placed symmetrically to the quartz tube. The peak wavelengths of the two lamps are 254nm and 365nm, respectively, and both are 8W with 600mm long. Aluminum foils are attached to the inner wall of the Perspex tube of 100mm inner diameter to diminish loss of the light radiation. The flow sheet of the tubular photoreactor is shown in Figure 4-14.

For formaldehyde, the formaldehyde concentration in the air stream is generated by changing formaldehyde concentration of formaldehyde aqueous solution from 1 to 10mg/ml, and the level of formaldehyde aqueous solution is below the tube for the air of route one. For other VOCs, their solution is put into the tube installed inside the VOC generator. Their concentration in the air stream is achieved by changing the nominal size of the tube for VOCs from 6 to 20mm.

Relative humidity is controlled by dividing the dry air into two routes. Route one is through the VOC generator, and route two bypasses the VOC generator and mixes with route one before flowing into the reactor. Relative humidity in a range of 8-100% is achieved in the present experiments.

After being dried by a bed of silicon gel, compressed air flows through a VOC generator before entering the reactor. The air is controlled by a mass flow controller, generating VOC loaded air stream. Air flow rates of 200 standard cubic centimeter per min (sccm) (for methanol) or 400sccm (for other VOCs) are used in the experiments, which give a residence times of about 6.8 or 3.4 seconds in the reactor, respectively, and Reynolds number about 37.5 and 75, respectively.

The air stream is sampled about every 30 minutes at both the inlet and outlet, and VOCs' concentration of the air stream is analyzed by an Agilent 6890N gas chromatograph. After the concentration at the inlet and outlet reaches equilibrium, the UV lamps are turned on to start degradation measurements.

Table 4.5. Carbon doped TiO_2 nanoparticles used in the experiments

Sample No.	Rutile mass fraction (-)	Carbon content % (wt)	Mean particle size (nm)
1	0.201	0.0	24
2	0.21	22.29	62
3	0.201	2.28	57
4	0.205	4.618	57
5	0.217	7.023	39

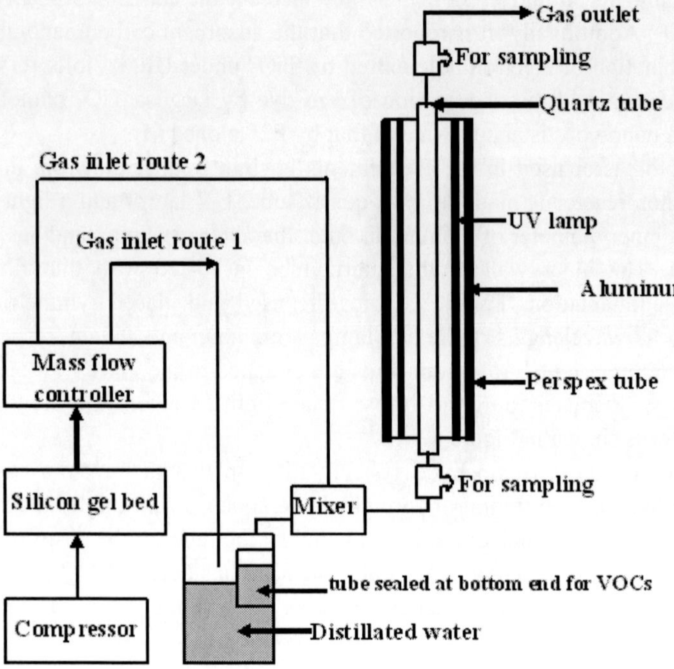

Figure 4.14. Illustration of tubular photocatalytic reactor for VOCs by C_TiO_2 nanoparticles.

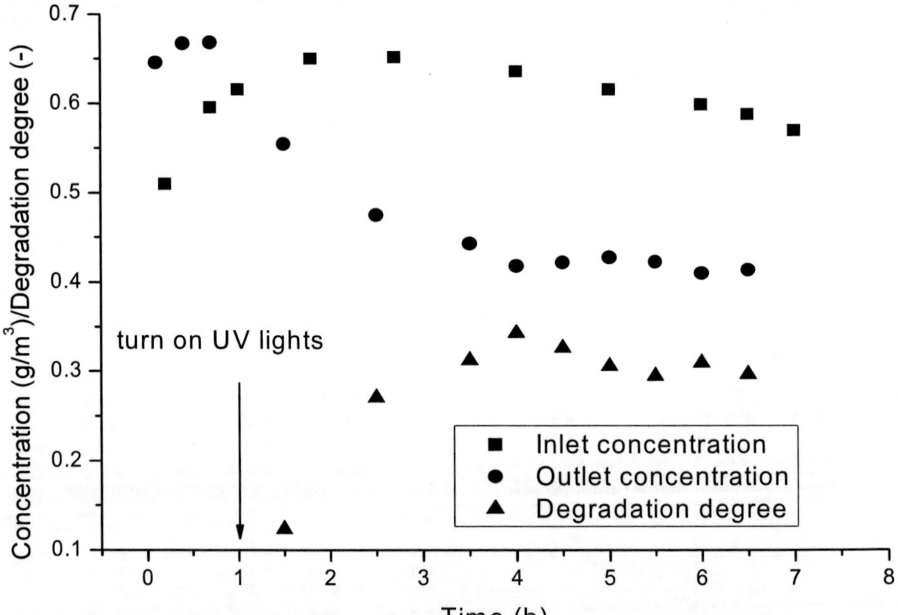

Figure 4.15. A typical measurement results of inlet and outlet concentrations *versus* time (formaldehyde).

Five carbon doped TiO_2 nanoparticles are used to carry out the photodegradation experiments. Their properties of rutile mass fraction, carbon content, mean particle size are tabulated in Table 4-5. Rutile mass fraction is about 0.2, mean particle size ranges from 24~62nm, carbon content is from 0~22.29% by weight.

Figure 4-15 is typical measurement results of inlet and outlet concentrations versus time for formaldehyde, the formaldehyde concentration in the air stream is about 0.684g/m^3, relative humidity is about 80% and three layers of TiO_2 thin films of sample No.4 (4.618% carbon content by weight) are used. The degradation degree could thus be obtained and is also plotted in Figure 4-5, and it could be seen that about 34% of the formaldehyde has been degraded in the experimental conditions.

Effect of number of layers of TiO_2 films on degradation degree is illustrated in Figure 4-16 for formaldehyde at a concentration in the air stream of about 0.684g/m^3 and 80% of relative humidity. Sample run No.4 is used in the test; its carbon content is 4.618% by weight. The figure shows that three layers of TiO_2 thin films have highest degradation degree. It has been discussed in section 4.2 that the effective thickness of three layers is about 99.6nm giving a catalyst loading of 38μg/cm^2. Optimum amount of catalyst was also observed for the photodegradation of propyzamide by a TiO_2 loaded with carbon black synthesized by sol-gel method, and it is the reflection of the utilization of TiO_2 particles in photoexcitation [12].

Effect of carbon contents on degradation degree is shown in Figure 4-17 for formaldehyde at the formaldehyde concentration in the air stream of about 0.684g/m^3, 80% of relative humidity and three-layers of TiO_2 thin films. It could be seen that sample No.4 with 4.618% carbon content in weight has highest degradation degree.

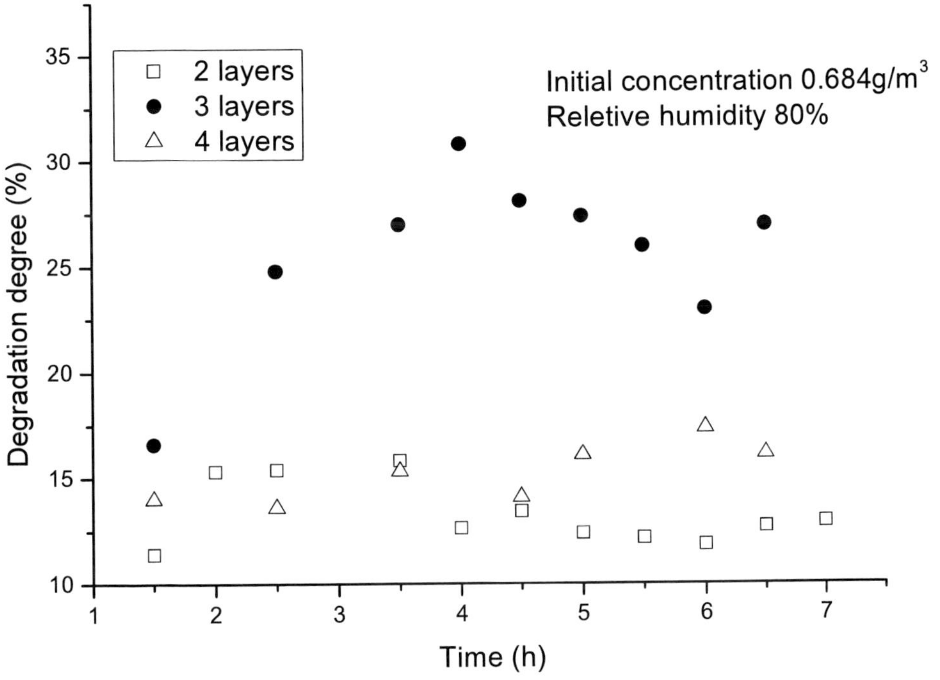

Figure 4.16. Effects of number of layers on degradation degree (formaldehyde).

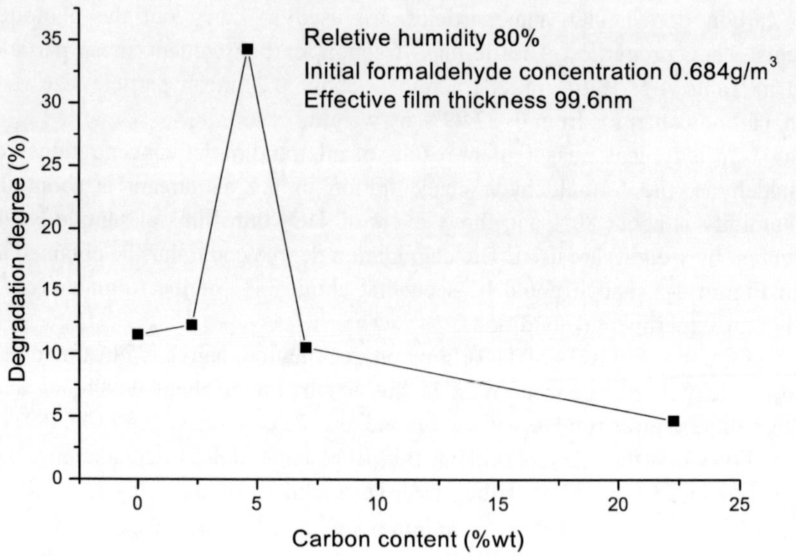

Figure 4.17. Effects of carbon content on degradation degree (formaldehyde).

Effects of relative humidity and initial VOCs' concentration in the air stream on degradation degree of different VOCs are shown in Figure 4-18 and 4-19, respectively, by using sample No.4 (4.618% carbon content in weight) with three layers of TiO_2 thin films. The figure reveals that 80% is the most favorable relative humidity for the photodegradation of formaldehyde and benzene, and 60% and 40% are for the photodegradation of toluene and methanol, respectively; and degradation degree increases rapidly as the concentration in the air stream decreases.

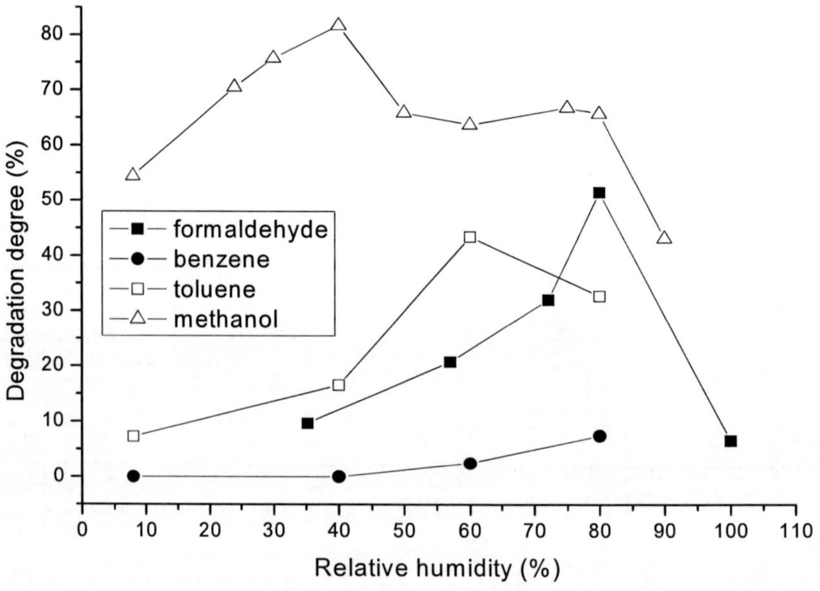

Figure 4.18. Effects of relative humidity on degradation degree.

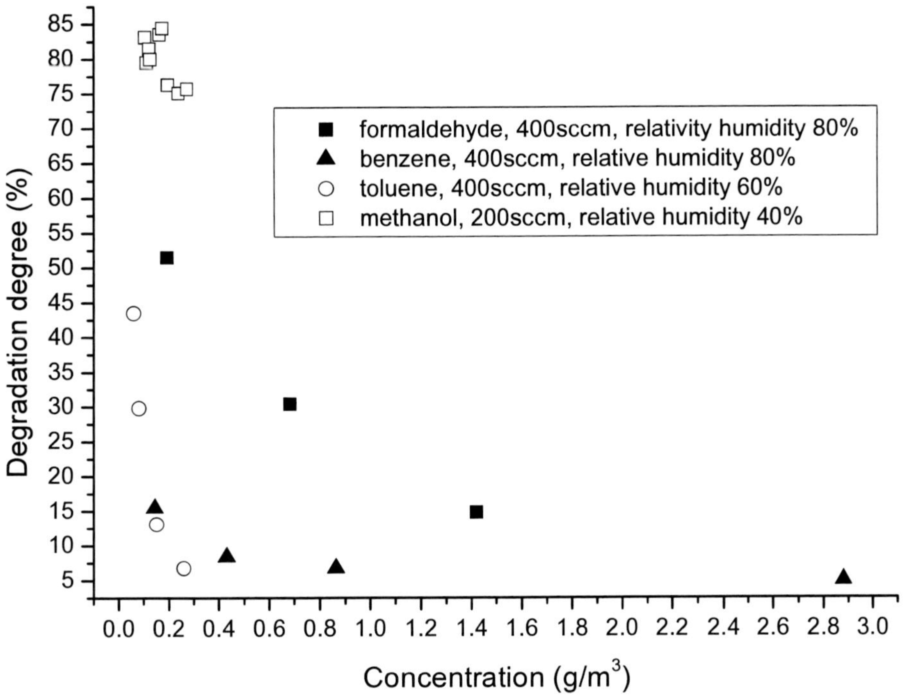

Figure 4.19. Effects of initial VOCs' concentration on degradation degree.

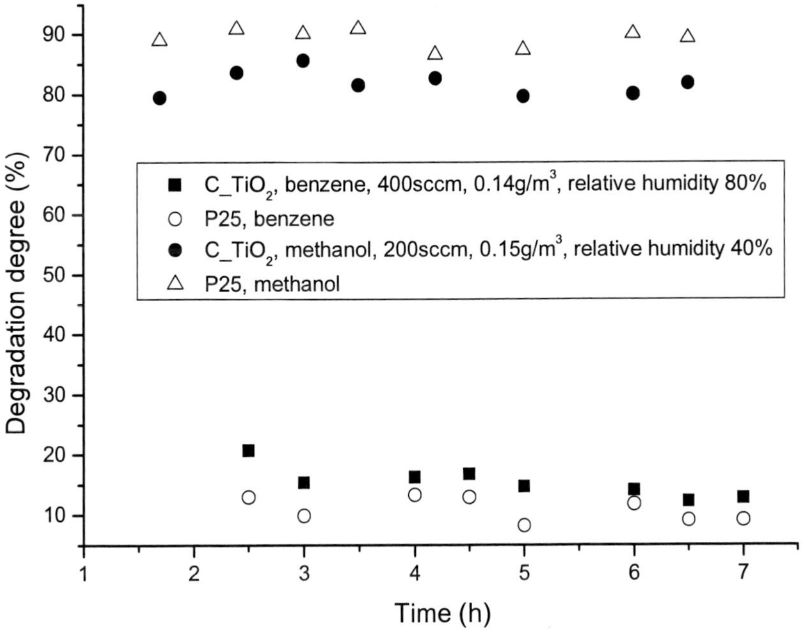

Figure 4.20. Comparisons on degradation degree between carbon doped TiO_2 and P25 photocatalyst.

It might mean that the concentration in the air stream exceeds the saturation behavior in the degradation of organic substrates [2], and the carbon doped TiO_2 is in favor of zero-emission applications in the treatment of environmental contaminations. Numerous studies indicate that a certain degree of humidity, usually exceeding the amount produced by the oxidation of organics is necessary to maintain the hydroxylation and to avoid the blockage of the TiO_2 surface by partially oxidized products [15-16].

Comparisons on degradation degree between carbon doped TiO_2 nanoparticles (sample No.4 of 4.618% carbon content by weight) and P25 photocatalyst are shown in Figure 4-20 for benzene and methanol. The results show that the photocatalytic activity of the carbon doped TiO_2 nanoparticles and P25 photocatalyst are comparative.

4.4. Photodegradation of VOCs and Ammonia by Mixed Phase TiO_2 Nanoparticles

Tubular photoreactors used in the experiments are shown in Figures 4-2 and 4-3 for VOCs and ammonia, respectively. The quartz tube has an inner diameter of 40mm, an outer diameter of 50mm, and has an effective length of 1200mm. Three layers of thin film of TiO_2 are coated on the inner wall of the quartz tube by sedimentation. The effective thickness of the TiO_2 films is about 99.6nm to give a catalyst loading of 38μg/cm^2. The peak wavelength of the UV lamp is 254nm and is 40W with 1200mm long to give 25mW/cm^2 UV light irradiation.

An 8-mm-i.d. tube is installed about 50mm below the lid of the VOC generator with an opening of area 18mm^2 located 30mm above the tube bottom keeping VOC level just below the opening inside the tube, as shown in Figure 4-2. The flow rate of air inlet route one is kept at 10 slm, which gives a residence time of the air stream in the photoreactor of about 5.7 s. The flow rate of air inlet route two ranges from 0.02~0.6 slm to give the inlet VOC concentration at ppb levels.

The air stream is sampled every 30 min at both the inlet and outlet by using Tenax tubes at a flow rate of 0.5 L/min for 1–5 min corresponding to the air flow rate via the VOC generator from 0.02 to 0.6 L/min, and VOC concentration of the air stream is analyzed by gas chromatography/mass spectrometry (GCMS, Shimadzu QP2010plus).

NH_3 flow rate is controlled by a mass flow controller in a range of 0-5sccm, giving the NH_3 initial concentration from 2-40mg/m^3. The air stream is sampled at both the inlet and outlet by adsorption using sulfuric acid solution of 0.01mol/L, and NH_3 concentration in the solution is analyzed by an ammonia-nitrogen analyzer, the conversion degree could then be obtained.

4.3.1. Photodegradation of Benzene and Toluene

Volatile organic compounds (VOCs) are widely used in industrial processes and domestic activities, leading to water and air pollution. Among the VOCs, benzene and toluene poses the highest cancer risk. One recent report demonstrated that exposure to even 1 ppm of benzene can reduce blood cell counts and cause hematotoxicity in factory workers [17].

Benzene and toluene are not only one of the main air pollutants of volatile organic compounds in industrial process and domestic activities, but also one of the main constituents in landfill gas [18]. Typical concentration of benzene in landfill gas in China is in about a hundred ppb level [18], it could be estimated that benzene emitted to atmosphere from landfills can be up to 10^3 m^3 annually.

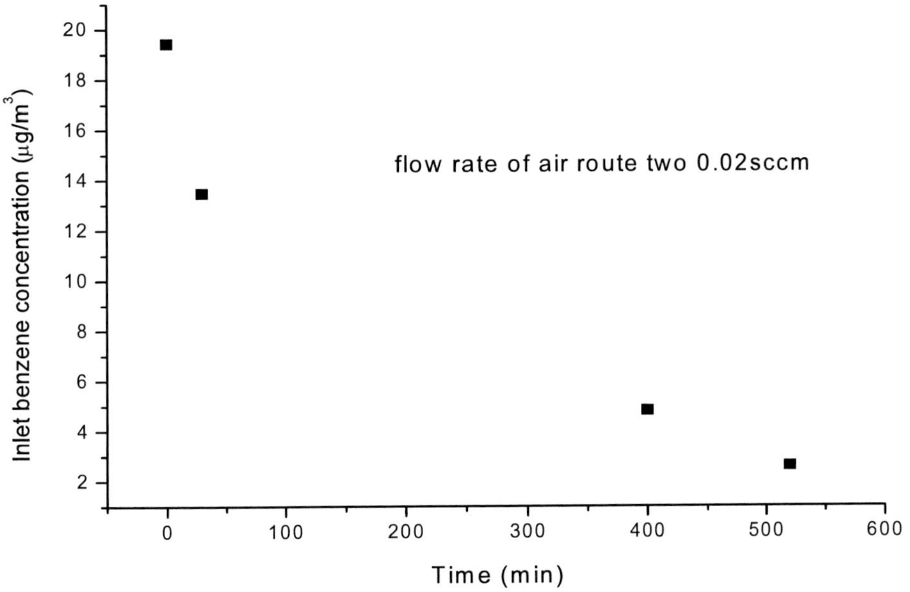

Figure 4.21. A typical changes of inlet benzene concentration *versus* time.

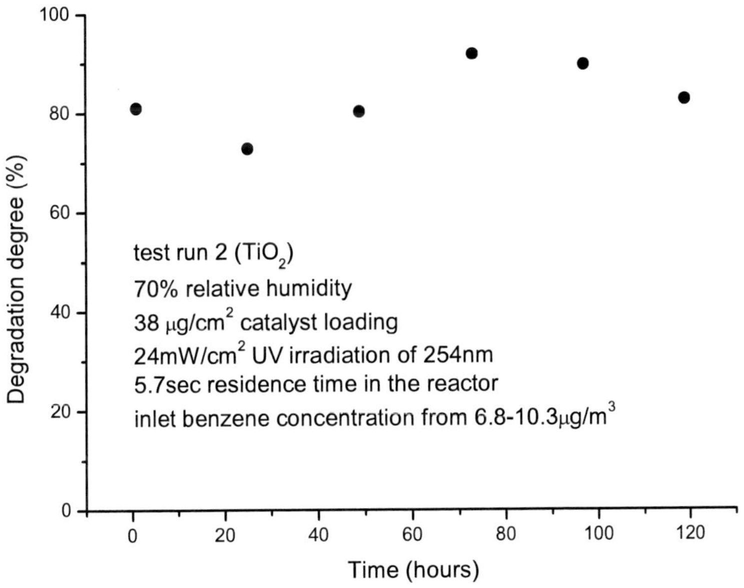

Figure 4.22. Time course of photodegradation of benzene.

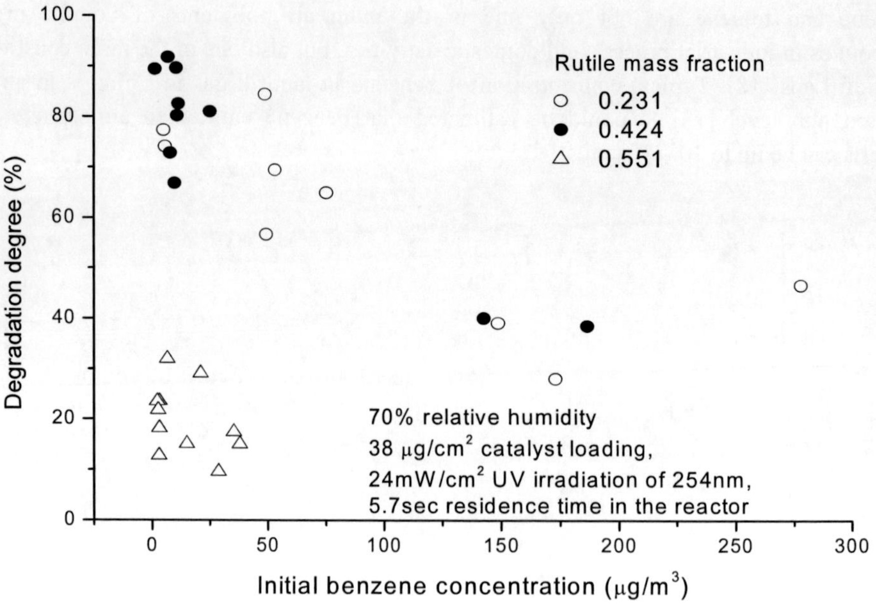

Figure 4.23. Effects of inlet benzene concentration and rutile mass fraction on degradation degree.

Because of evaporation, the inlet VOC concentration decreases with time, as shown in Figure 4-21 for benzene, typically from about 20 to about 2μg/m^3 in about 10 h, thus calling for benzene refill every 3 h. Since the residence time of the air stream in the photoreactor is only about 5.7 s, and sampling lasts some 1–5 min, the change of the inlet benzene concentration within the sampling period is minimal, e.g. <10% by least square analysis. The inlet VOC concentration is thus taken as the instantaneous value at the time of sampling.

Contrast experiments are carried out under the conditions of with illumination but no catalyst, and with TiO_2 but no illumination. The results show that the concentration of benzene or toluene (100μg/m^3) changed little under the conditions of both with illumination but no catalyst, and with TiO_2 but no illumination. The blank tests proved that benzene and toluene is kinetically stable without illumination and photocatalyst. It is also suggested that dark adsorption–desorption for benzene and toluene from gas phase is small; and the adsorption equilibrium could be established quickly.

Figure 4-22 is the time course of photodegradation of benzene at the inlet benzene concentration from 6.8-10.3μg/m^3, under 70% relative humidity. In order to maintain stable inlet benzene concentration, apart from refilling into the tube for benzene in the benzene generator about every 3 hours, benzene is refilled again about 30 minutes before sampling. The conversion is between 80-90% within the test time of about 120 hours. That deactivation caused by the accumulation of the polymerized by-products on catalyst surface as reported by Einaga, Futamura and Ibusuki is not observed under the present experimental conditions [19,20].

Effects of inlet benzene concentration and rutile mass fraction on degradation degree are shown in Figure 4-23 under 70% relative humidity. The figure shows that photocatalytic activity is higher when rutile mass fraction is a range of 0.2 to 0.4. At the inlet benzene concentration of about 10μg/m^3, 90% conversion is achieved and degradation degree decreases to about 40% as the inlet benzene concentration increases up to about 300μg/m^3.

Effects of initial toluene concentration and rutile mass fraction on degradation degree are shown in Figure 4-24 under 70% relative humidity. Again, the figure shows that photocatalytic activity is higher when rutile mass fraction is in between 0.2 to 0.4; and degradation degree increases rapidly when the initial concentration in the air stream is less than 300μg/m^3. At the initial toluene concentration of about 50μg/m^3, 90% conversion is achieved and degradation degree decreases to about 50% as the initial toluene concentration increases up to about 300μg/m^3.

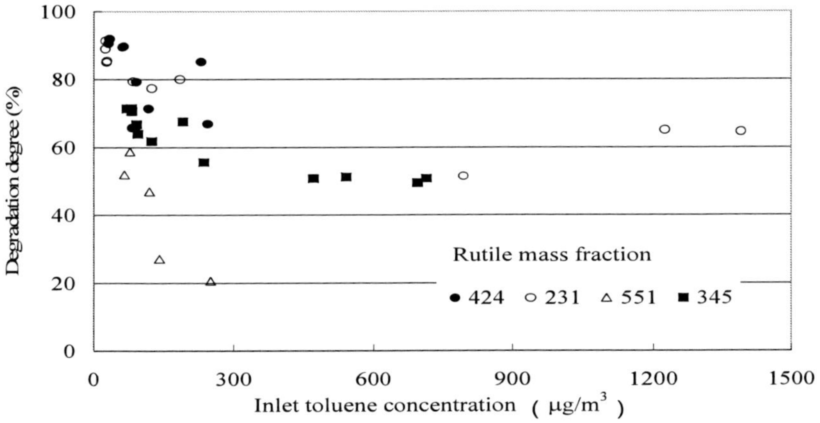

Figure 4.24. Effects of initial toluene concentration and rutile mass fraction on degradation degree.

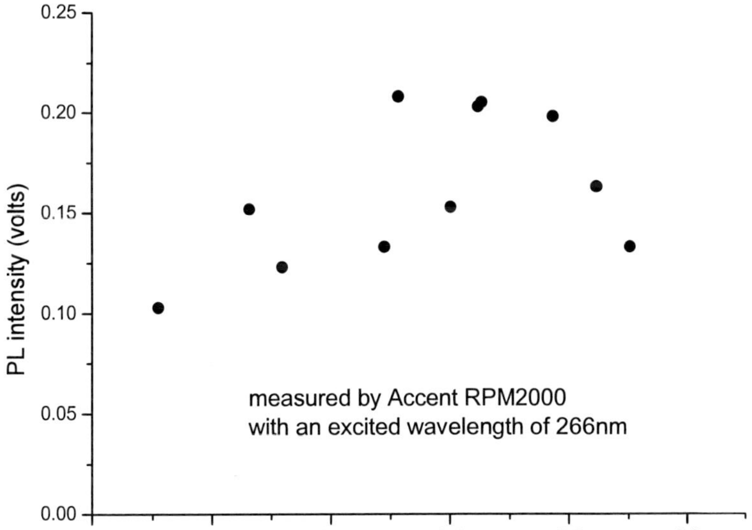

Figure 4.25. PL intensity *versus* rutile mass fraction.

Correlations of rutile fraction and the PL intensity at the wavelength of 400–420 nm of the TiO_2 nanoparticles are plotted in Figure 4-25. It could be seen that PL intensity increases as rutile mass fraction increases up to about 0.4, and decreases as rutile fraction above 0.4, reflecting electrons from valence to conductive bands more vigorously, and electronic coupling in the contacts between anatase and rutile phases [21]. The results that photocatalytic activity is higher when rutile mass fraction is between 0.2 to 0.4, yield that anatase-rutile interface prohibits the recombination of photoinduced electron (e^-) and hole (h^+) pairs.

4.3.2. Photodegradation of Ammonia

Ammonia (NH_3) is a major contributor to secondary aerosol formation in the atmosphere as it reacts rapidly with sulfuric (H_2SO_4), nitric (HNO_3) and hydrochloric (HCl) acids to form ammonium (NH_4^+) aerosol, and ammonium aerosol affects Earth's radiative balance, both directly by scattering incoming radiation and indirectly by acting as cloud condensation nuclei [22]. Recently it was found that for a 50% reduction NH_3 availability, inorganic $PM_{2.5}$ was reduced by 29% in western Pennsylvania of USA [23]. Additionally, NH_3 in atmosphere will affects the nitrogen status of terrestrial and aquatic ecosystems, as atmospheric deposition of reduced nitrogen (i.e. NH_3+NH_4^+) to terrestrial and aquatic ecosystems may exacerbate eutrophication of aquatic ecosystems [24] and may lead to soil acidification, decreased resistances to climatic stressors, a significant shift in the nutrient economy and shifts in species composition.

A recent survey carried out by the State Environmental Protection Administration of China reveals that landfills account for 88% of all urban waste management facilities and accommodate 94% of the urban waste collected [25]. Typical concentration of NH_3 in landfill gas in China is in a few ppm level [18], it could be estimated that NH_3 emitted to atmosphere from landfills can be up to 10^5 m^3 annually. Removal of such NH_3 has significant importance for both clean atmosphere and the nitrogen balance in terrestrial and aquatic ecosystems.

Great efforts have been devoted to the removal of NH_3 from wastewater, however, it lacks of reports of NH_3 removal directly from gas phase and such efforts have significant importance in removing NH_3 from landfill gas. Photocatalytic oxidation offers potentially a facial and economical method for removal inorganic and organic pollutants from waste waters. TiO_2 is currently the best known and most widely used highly efficient photocatalytic material because it is stable and has higher light energy conversion efficiency due to its suitable band gap positions.

Effects of initial NH_3 concentration on degradation degree are shown in Figure 4-26. The figure shows that degradation degree increases rapidly as the initial NH_3 concentration decreases. At the initial NH_3 concentration of about 5mg/m^3, 60% conversion is achieved and degradation degree is less than 20% when the initial NH_3 concentration is above 20mg/m^3. Same dependence of degradation degree on initial concentration has been discussed in sections 4-1 and 4-2.

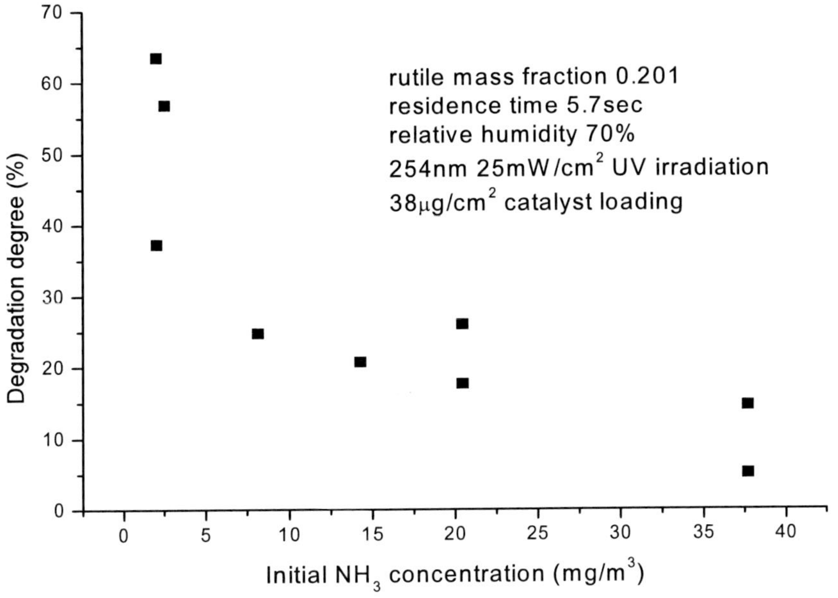

Figure 4.26. Effects of initial NH_3 concentration.

Effects of relative humidity and rutile mass fraction on degradation degree are shown in Figure 4-27. Increasing degradation degree with relative humidity from 30-70% implies that oxidation reactions of H_2O to ·OH hydroxyls on the valence band of the TiO_2 dominate the photodegradation reactions [2]. The figure shows that photocatalytic activity is higher when rutile mass fraction is between 0.2 to 0.4, yielding that anatase-rutile interface prohibits the recombination of photoinduced electron (e^-) and hole (h^+) pairs, the same dependence on rutile fraction as that for degradation of benzene and toluene.

4.3.3. Separation of Photon-Induced e^-/h^+ Pairs by Rutile Phase

Gratzel and Frank [26] has demonstrated that photoinduced electron (e^-) and hole (h^+) pairs migrate from interior to the surface much faster than bulk recombination of electron (e^-) and hole (h^+) pairs. And surface recombination process is in about a few nanosecond scale [27], which is faster or is compete to the oxidation characteristic time of 10 nanoseconds on valence band of semiconductor oxides [28]. Hence separation of photoinduced electron (e^-) and hole (h^+) pairs on catalyst surface will play crucial roles in determining photocatalytic activities.

Rutile and anatase phases are semiconductors with a band gap of 3.2 eV for anatase and 3.1 eV for rutile. The conduction band (CB) edges are about -0.5eV for anatase and -0.4eV for rutile versus normal hydrogen electrode (NHE) at pH 7, respectively [29]. Thus, photo-induced (e^-) in anatase phase will migrate to rutile surface due to that the potential of conductive band of rutile is lower than that of anatase. Consequently, more holes are ready on anatase surface to cause oxidation reactions, whereas electrons are probably consumed for the reduction of O_2 on rutile surface as illustrated in Figure 4-28.

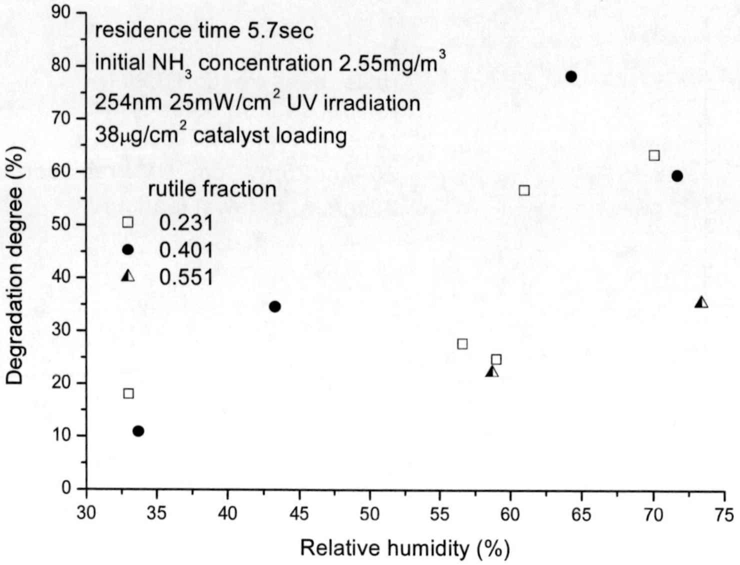

Figure 4.27. Effects of relative humidity and rutile fraction.

This mechanism is reported for anatase TiO_2 and rutile SnO_2 interface as the CB edges of rutile SnO_2 is about +0.07eV versus normal hydrogen electrode (NHE) at pH 7 [30,31], lower than that of anatase TiO_2.

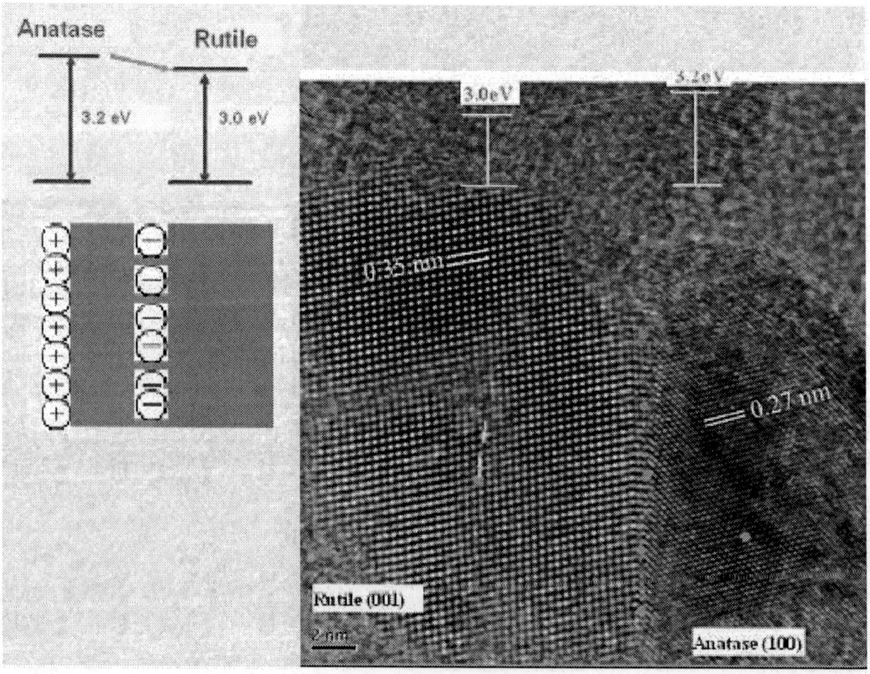

Figure 4.28. Illustration of the separation of photon-induced e^-/h^+ pairs by rutile phase.

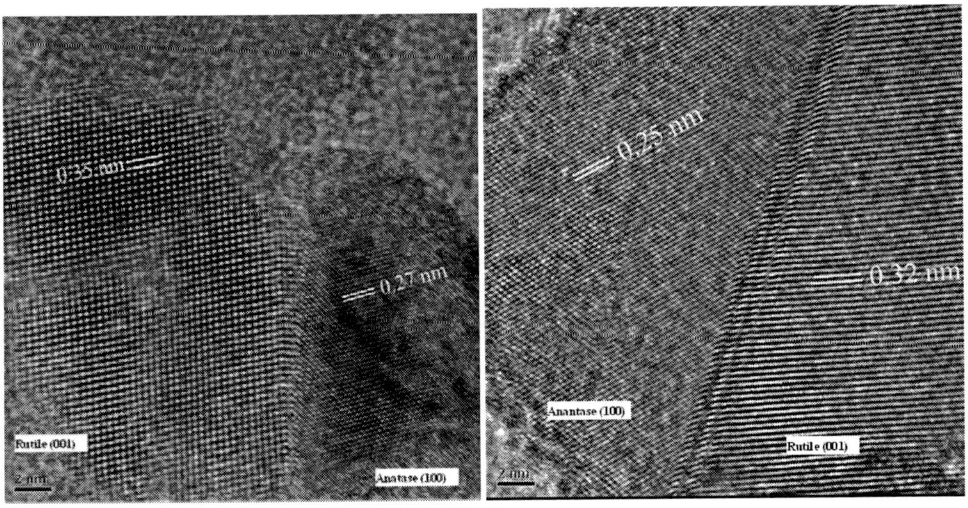

Figure 4.29. Typical anatase/rutile phases on the TiO_2 nanoparticle surface.

Microwave absorption experiments using time resolved microwave conductivity (TRMC) have shown that rutile phase can help to avoid the recombination of photon-induced e^-/h^+ pairs [32]. Photodegradation experiments with mixed phase titania showed that rutile phase is related to the photo-induced e^-/h^+ life-time as estimated qualitatively via photoluminescence (PL) analysis [21].

Experimental results that photocatalytic activity is higher when rutile mass fraction is in between 0.2 to 0.4, also demonstrate that rutile phase can help to avoid the bulk recombination of photon-induced e^-/h^+ pairs, thus more photon-induced e^-/h^+ pairs migrate to the catalyst surface, corresponding higher PL intensity.

Figure 4-29 shows typical anatase/rutile phases detected by HRTEM on the TiO_2 nanoparticle surface. The lattice fringes are of around 0.33-0.35nm, corresponding to (001) planes of rutile TiO_2, and the lattice fringes are of 0.25-0.27nm, corresponding to the (100) planes of anatase TiO_2.

4.4. Photodegradation of Benzene by SnO_2 Nanoparticles

Since TiO_2-based photocatalysts have low adsorption capacities, the ·OH radical mechanism in gas phase is sensitive to relative humidity [33]. In the absence of water vapor, the photocatalytic degradation of formaldehyde [34], acetone [35] toluene [36,37] is seriously retarded, which may imply that direct hole oxidation in the gas phase is insignificant with TiO_2-based photocatalysts. On the other hand, excessive water vapor on the catalyst surface will inhibit the reaction rates because the presence of water vapor competes with pollutants for adsorption sites, thus reducing the pollutant removal rate [2,33 -35,38].

Process economics estimated by Miller and Fox [39] indicated that photocatalytic treatment of lightly contaminated air is commercially attractive only for conversions of high photoefficiency reactants, and air contaminants exhibiting lower apparent photoefficiency (below 20–30%) were found to compare less favorably to incinerative and carbon adsorption processes [39]. Sensitivity to relative humidity will make TiO_2-based PCO processes in low performance in gas purification applications because in most environmental applications relative humidity is subject to change with climate and weather conditions and/or with operational conditions of a particular process. Exploring humidity-tolerant photocatalyst or photocatalytic mechanisms therefore has great significance both scientifically and practically, and direct hole oxidation could be a viable and efficient method to meet the challenge of relative humidity variations in environmental applications.

Like TiO_2, SnO_2 is also an n-type semiconductor and has been used intensively for gas sensor, transparent conductor, nanoeletronic device and oxidation catalyst [40]. Photocatalytic studies on TiO_2/SnO_2 systems have been reported and the system enhances photocatalytic activity by charge separation mechanism, as the CB edges of rutile SnO_2 is about +0.0eV versus normal hydrogen electrode (NHE) at pH 7, lower than that of TiO_2, and the VB edges of rutile SnO_2 is about 3.6eV versus normal hydrogen electrode (NHE) at pH 7, higher than that of TiO_2 [30,31,41]. Recent results in photodegradation of methylene blue (MB) has shown that rutile SnO_2 made by thermal decomposition of tin chloride pentahydrate ($SnCl_4{\cdot}5H_2O$) has similar catalytic activity with TiO_2 phototcatalyst (P25) [42].

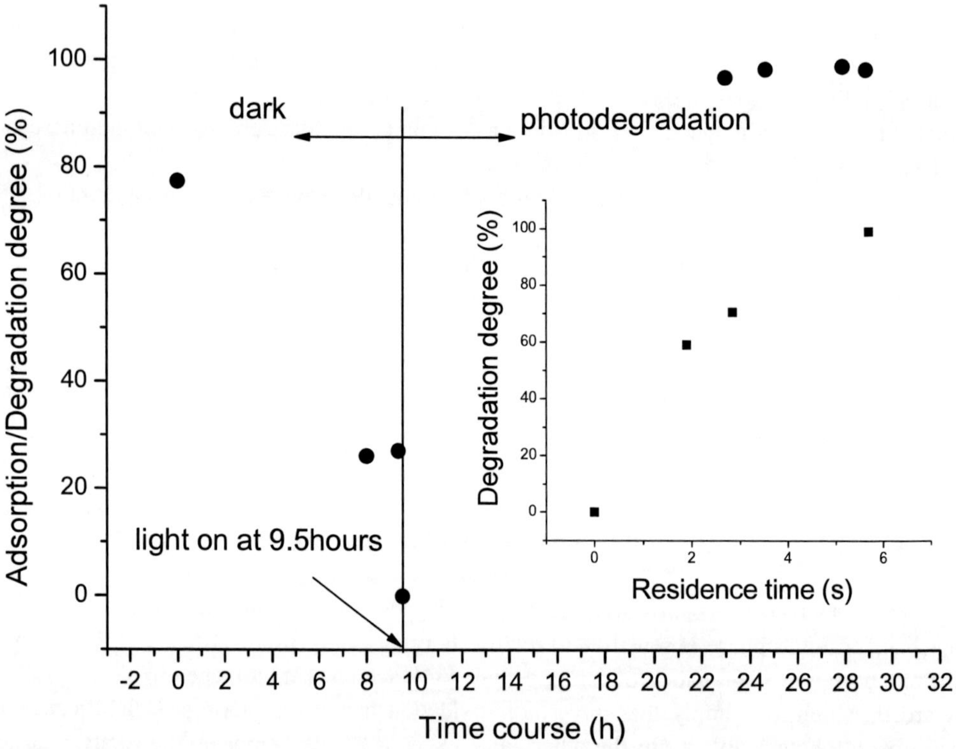

Figure 4.30. Time course of adsorption/degradation of benzene (the insert illustrated degradation degree versus residence time).

The time course of degree of adsorption/degradation is shown in Figure 4-30 in humid air with a relative humidity of about 25%, benzene inlet concentration of about $100\mu g/m^3$. It shows that equilibrium between inlet and outlet of the photoreactor is reached in about 9.5 hours, much longer than that for TiO_2 nanoparticles, in which equilibrium is reached in minutes [43]. After the light is turned on, benzene is almost completely degraded for a long period of time. That deactivation caused by the accumulation of the polymerized by-products on catalyst surface as reported by Einaga et al.[19,20] is not observed in the present experiments. The insert in Figure 4-30 shows that the degradation degree is nearly proportional to the residence time of benzene-loaded air inside the photoreactor, further verifying the viability of the reactor system.

Degrees of degradation versus benzene inlet concentration in humid air (25% relative humidity), dry air (bottle air) and N_2 are shown in Figure 4-31. It can be seen that photocatalytic activities in dry air and N_2 are only slightly lower than those in humid air. The insert of Figure 4-31 shows the time course of degree of degradation, that phtotdegradation in dry air is in the first 3 hours, then is in humid air (25% relative humidity) for 18 hours (the light on), and is then in N_2 for 3 hours. Photocatalytic degradation seriously retarded in the absence of water vapor [34-37] is not detected. In dry air and N_2, the ·OH radical mechanism could be eliminated as no water molecule is available. Because the CB edges of rutile SnO_2 is about +0.0eV versus normal hydrogen electrode (NHE) at pH 7, rutile SnO_2 can not efficiently participate in the reduction of O_2 (the CB edges of O_2, is -0.3eV versus normal hydrogen electrode (NHE) at pH 7) [41]. Direct hole oxidation therefore must play a vital role since 3.6eV of VB edges of rutile SnO_2 is much higher than the oxidation potential of aromatic hydrocarbons [2], together with the Mars-van Krevelen mechanism, in which benzene is oxidized by consuming the lattice oxygen of SnO_2, which in turn is re-oxidized by gas-phase oxygen [40]. The degree of degradation in humid air (25% relative humidity) is slightly higher than those in dry air and N_2 is due to the contribution by the ·OH radical mechanism, again indicating the ·OH radical mechanism plays a minor role.

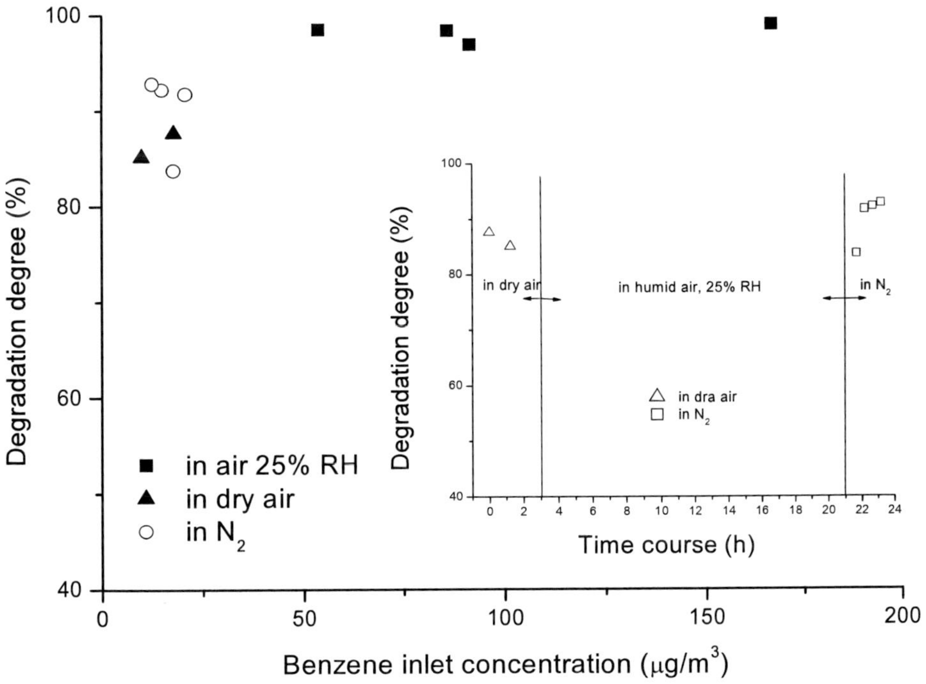

Figure 4.31. Photodegradation of benzene in humid air, dry air and N_2.

4.5. Apparent 1st Order Rate Constant

The diffusion equation in axi-symmetric cylindrical co-ordinates by neglecting axial diffusion and at the laminar flow regime could be given by [44]

$$\frac{\partial Y}{\partial \eta}=\frac{D_A \tau}{2R^2}\left(1-\xi^2\right)^{-1}\left(\frac{1}{\xi}\frac{\partial Y}{\partial \xi}+\frac{\partial^2 Y}{\partial \xi^2}\right) \tag{4-12}$$

where Y is the normalized concentration c/c_{in}, η is the normalized axial distance z/L, ξ is the normalized radial distance r/R, D_A is the diffusion coefficient in air, τ is the gas residence time inside the reactor, R and L are the inner diameter and effective length of the reactor, respectively, and c_{in} is the concentration at the inlet.

By using 2nd order central differential scheme in radial direction, it can be discredited into

$$\frac{dY(i,\eta)}{d\eta}=A(i)Y(i+1,\eta)+B(i)Y(i,\eta)+C(i)Y(i-1,\eta) \tag{4-13}$$

where

$$\left.\begin{array}{l} A=\beta\left[1/(2\xi\Delta\xi)+1/\Delta\xi^2\right]/\left[2(1-\xi^2)\right] \\ B=\beta\left(-1/\Delta\xi^2\right)/\left[(1-\xi^2)\right] \\ C=\beta\left[-1/(2\xi\Delta\xi)+1/\Delta\xi^2\right]/\left[2(1-\xi^2)\right] \end{array}\right\} \tag{4-14}$$

$\Delta\xi$ is the normalized radial step length, and $\beta=D_A\tau/R^2$.

From symmetrical condition at the central axis, e.g. $\xi=0, Y(1,\eta)=Y(-1,\eta)$, it can be obtained that

$$\frac{dY_0}{d\eta}=A_0 Y(1,\eta)-B_0 Y(0,\eta) \tag{4-15}$$

$$\left.\begin{array}{l} A_0=\beta/\Delta\xi^2 \\ B_0=-\beta/\Delta\xi^2 \end{array}\right\} \tag{4-16}$$

From the condition $-D_A\frac{\partial c}{\partial r}\cdot 2\pi R dz=\dot{r}2\pi R dz$ on the wall $(\xi=1)$ and 2nd order upwind differential scheme, it can be obtained that

$$Y_I=-\frac{2}{3}\alpha\Delta\xi+\frac{4}{3}Y_{I-1}-\frac{1}{3}Y_{I-2} \tag{4-17}$$

where

$$\alpha = \frac{\dot{r}}{D_A c_{in}/R} \tag{4-18}$$

where $\dot{r}$ is the degradation rate and for 1st order reaction, the degradation rate is expressed to

$$\dot{r} = kc \tag{4-19}$$

where k is the 1st order degradation rate constant in the unit of m/s.

Thus, the following 1st order ordinary differential equations can be obtained

$$\begin{cases} \dfrac{dY_0}{d\eta} = A_0 Y(1,\eta) - B_0 Y(0,\eta) \\ \cdots\cdots \\ \dfrac{dY_{i-1}}{d\eta} = A(i-1)Y(i,\eta) + B(i-1)Y(i-1,\eta) + C(i-1)Y(i-2,\eta) \\ \cdots\cdots \\ Y_I = -\dfrac{2}{3}\alpha\Delta\xi + \dfrac{4}{3}Y_{I-1} - \dfrac{1}{3}Y_{I-2} \end{cases} \tag{4-20}$$

In the calculations, normalized radial step length, $\Delta\xi$, is taken as 10^{-2}. From the numerical stability of equations (4-20) [44], normalized axial step length, $\Delta\eta$, is taken as 10^{-7}. For a given value of α, equations (4-20) are integrated by the 4th order Runge-Kutta method, the concentration at the outlet could be then obtained. The scheme is repeated until the calculated concentration at the outlet for a given value of α is the same as that measured. The 1st order degradation rate constant k could be then obtained by the α value from equations (4-18) and (4-19).

Experiments of degradation of formaldehyde by carbon doped TiO_2 nanoparticles are performed by using a tubular photoreactor as shown in Figure 4-14. Air flow rate of 400sccm is used in the experiments, which give a residence time of about 3.4 seconds in the reactor and Reynolds number about 75. The formaldehyde concentration of the formaldehyde aqueous solution is 3mg/ml, which generates about 570ppm of formaldehyde loaded air stream. The relative humidity of the formaldehyde loaded air stream keeps about 80%, which is the optimum relative humidity for photodegradation of formaldehyde. The properties of carbon doped TiO_2 nanoparticles and degradation degree are given in Table 4-6.

For photodegradation of formaldehyde by titania, the photocatalytic rates could be well described by the Langmuir–Hinshelwood model [44]. But, the apparent first-order reaction rate constant is useful in evaluating the photocatalytic capacity of titania photocatalysts and in modeling the performance of photocatalytic reactors [45-47], in particular, under low pollutant concentrations.

The 1st order rate constant k obtained for the samples is also given in Table 4-6 and is in a range of $1.37 \cdot 10^{-4}$m/s to $1.17 \cdot 10^{-3}$m/s. In the calculations, the diffusion coefficient of formaldehyde in air D_A is taken as 1.335×10^{-5}m^2/s.

Table 4.6. Physical properties of carbon doped TiO_2 nanoparticles and apparent 1st order rate constant of photodegradation of formaldehyde

Sample No.	Mean particle size (nm)	Rutile mass fraction (-)	Carbon content % (wt)	Degradation degree (%)	1st order rate constant (m/s)
1	24.3	0.110	0.0	16.06	6.35×10^{-4}
2	16.6	0.168	0.0	17.02	6.67×10^{-4}
3	32.3	0.166	1.324	11.80	4.53×10^{-4}
4	42.9	0.298	0.828	24.57	1.01×10^{-3}
5	23.6	0.201	0.0	11.55	4.41×10^{-4}
6	23.5	0.175	1.358	8.65	3.29×10^{-4}
7	16.6	0.119	2.719	8.56	3.21×10^{-4}
8	24.8	0.087	0.0	4.09	1.49×10^{-4}
9	26.0	0.126	0.0	6.71	2.57×10^{-4}
10	16.5	0.076	0.459	3.90	1.37×10^{-4}
11	61.9	0.21	22.29	4.69	1.72×10^{-4}
12	56.7	0.201	2.28	12.31	4.79×10^{-4}
13	56.5	0.205	4.618	27.73	1.17×10^{-3}
14	48.9	0.309	8.665	6.23	2.32×10^{-4}
15	37.7	0.275	10.083	13.63	5.30×10^{-4}
16	39.3	0.217	7.023	10.51	4.02×10^{-4}

Figure 4-32 plots the 1st order rate constant against rutile mass fraction for the samples with negligible amount of carbon content. It could be seen that the 1st order rate constant increases with increasing rutile mass fraction up to 0.3. In the synthesis of carbon doped TiO_2 nanoparticles, laminar diffusion flame CVD method is used as discussed in Chapter 1.

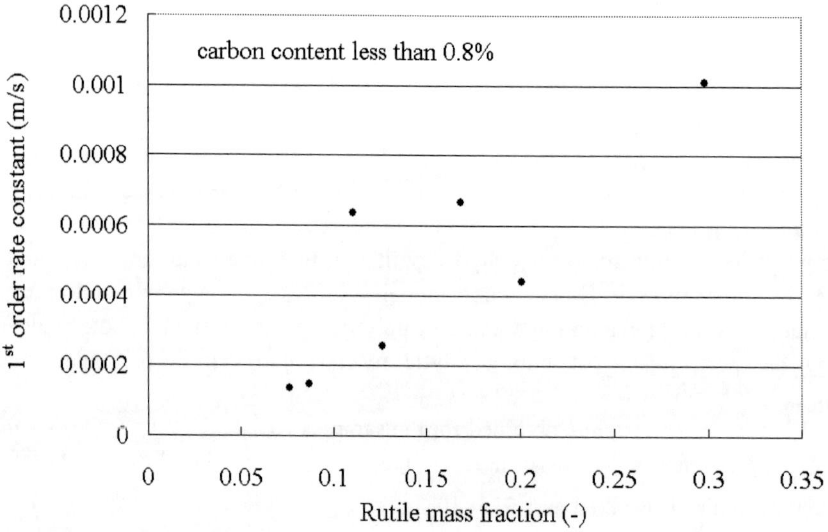

Figure 4.32. 1st order rate constant against rutile mass fraction.

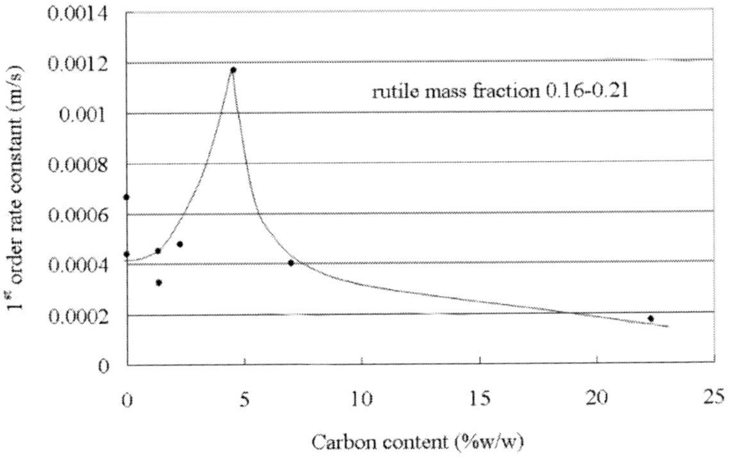

Figure 4.33. Effects of carbon content on the 1st order rate constant..

Rutil mass fraction is limited since temperature is lower in laminar diffusion flame CVD process.

Optimum rutile mass fraction of about 0.226 has been reported for photodegradation of 4-chlorophenol in aqueous suspension by a TiO_2, which was synthesized by homogeneous precipitation of aqueous solutions containing $TiOSO_4$ with urea [48], and optimum rutile mass fraction of about 0.3 has been reported in photodegradation of aniline by PhS-sensitied TiO_2 film [49].

It is discussed in the section 4.3 that photodegradation photocatalytic activity is higher when rutile mass fraction is between 0.2 to 0.4 for mixed phase TiO_2 by turbulent diffusion flame CVD method, demonstrates that rutile phase can help to avoid the bulk recombination of photon-induced e^-/h^+ pairs, thus more photon-induced e^-/h^+ pairs migrate to the catalyst surface, corresponding higher PL intensity.

Figure 4-33 shows the effects of carbon content on the 1st order rate constant for the test runs having rutile mass fraction from 0.16 to 0.21.

It shows again that the degradation degree occurs a maximum at the carbon content of about 5% by weight, and the degradation degree at the carbon content of about 5% is about 2.5 times of that at carbon content about zero or above 10%.

For the photodegradation of propyzamide by a TiO_2 loaded with carbon black synthesized by sol-gel method, optimum carbon black has been reported for about 20% by weight [12].

REFERENCES

[1] Brillas E., Mur E., Sauleda R., Sànchez L., Peral J., Domènech X. and Casado J. (1998). Aniline mineralization by AOP's: anodic oxidation, photocatalysis, electro-Fenton and photoelectro-Fenton processes. *Appl. Catal. B Environ.* 16, 31.

[2] Carp O, Huisman CL, Reller A (2004). Photoinduced reactivity of titanium dioxide. *Prog. Solid State Chem.* 32, 33.

[3] Frank S.N. and Bard A.J. (1977). Heterogeneous photocatalytic oxidation of cyanide ion in aqueous solutions at titanium dioxide powder. *J. Am. Chem. Soc.* 99, 303.

[4] Frank S.N. and Bard A.J. (1977). Heterogeneous photocatalytic oxidation of cyanide and sulfite in aqueous solutions at semiconductor powders. *J. Phys. Chem.* 81, 1484.

[5] Alberici R.M. and Jardim W.E. (1997). Photocatalytic destruction of VOCs in the gas-phase using titanium dioxide. *Appl. Catal. B Environ.* 14, 55.

[6] Fujishima A., Rao T.N. and Tryk D.A. (2001). Titanium dioxide photocatalysis. *J. Photochem. Photobiol. C*, 1, 1.

[7] Vorontsov A. V., Savinov E. N. and Jin Z. S. (1999). Influence of the form of photodeposited platinum on titania upon its photocatalytic activity in CO and acetone oxidation. *J. Photochem. Photobiol. A: Chem.* 125, 113.

[8] Palmisano L., Schiavello M., Sclafani A., Martin C., Martin I. and Rives V., (1994). Surface properties of iron-titania photocatalysts employed for 4-nitrophenol photodegradation in aqueous TiO_2 dispersion. *Catal. Lett.* 24, 303.

[9] Yu J. C., Xie Y., Tang H. Y., Zhang L., Chen H. C. and Zhao J. J. (2003). Visible light-assisted bactericidal effect of metalphthalocyanine-sensitized titanium dioxide films. *J. Photochem. Photobiol. A: Chem.* 156, 235.

[10] Qourzal S., Assabbane A., and Ait-Ichou Y. (2004). Synthesis of TiO_2 via hydrolysis of titanium tetraisopropoxide and its photocatalytic activity on a suspended mixture with activated carbon in the degradation of 2-naphthol. *J. Photochem. and Photobio. A: Chem.* 163, 317.

[11] Torimoto T., Okawa Y., Takeda N. and Yoneyama H. (1997). Effect of activated carbon content in TiO_2-loaded activated carbon on photodegradation behaviors of dichloromethane. *J. Photochem. and Photobio. A: Chem.* 103, 153.

[12] Takeda N., Iwata N., Torimoto T., and Yoneyama H. (1998). Influence of carbon black as an adsorbent used in TiO_2 photocatalyst films on photodegradation behaviors of propyzamide. *J. Catal.* 177, 240.

[13] Tian D., Yu G., and Zhang P.Y. (2002). Photocatalytic degradation of toluene in the gas phase under carbon black modified titanium dioxide. *J. Solar Energy* 23, 1.

[14] Yu Y., Yu J. C., Chan C. Y., Che Y. K., Zhao J. C., Ding L., Ge W. K. and Wong P. K. (2005). Enhancement of adsorption and photocatalytic activity of TiO_2 by using carbon nanotubes for the treatment of azo dye. *Appl. Catal. B: Environ.* 61, 1.

[15] Ameen M. M. and Raupp G. B. (1999). Reversible Catalyst Deactivation in the Photocatalytic Oxidation of Diluteo-Xylene in Air. *J. Catal.* 184, 112.

[16] Maira A. J., Yeung K. L., Soria J., Coronado J. M., Belver C. and Lee C. Y. (2001). Gas-phase photo-oxidation of toluene using nanometer-size TiO_2 catalysts. *Appl. Catal. B: Environ.* 29, 327.

[17] Lan Q., Zhang L.P., Li G.L., Vermeulen R. et al. (2004). Hematotoxicity in Workers Exposed to Low Levels of Benzene. *Science* 306, 1774.

[18] Xie X.Y. (2008). Final Report of Air Pollutants from Landfills on the east coast areas of Pudong of Shanghai.

[19] Einaga H., Futamura S. and Ibusuki T. (1999). Photocatalytic decomposition of benzene over TiO_2 in a humidified air stream. *Phys. Chem. Chem. Phys.* 1, 4903.

[20] Einaga H., Futamura S. and Ibusuki T. (2002). Heterogeneous photocatalytic oxidation of benzene, toluene, cyclohexene and cyclohexane in humidified air: comparison of

decomposition behavior on photoirradiated TiO_2 catalyst. *Appl. Catal. B: Environ.* 38, 215.

[21] Baiju K.V., Zachariah A. and Shukla S. (2009). Correlating Photoluminescene and Photocatalytic Activity of Mixed-phase Nanocrystalline Titania. *Catal. Lett.*, 130, 130.

[22] Yu F. (2006). Effect of ammonia on new particle formation: A kinetic H_2SO_4-H_2O-NH_3 nucleation model constrained by laboratory measurements. *J. Geophys. Res.* 111, D01204.

[23] Vayenas D.V., Takahama S., Davidson C.I. and Pandis S.N. (2005). Simulation of the thermodynamics and removal processes in the sulfate-ammonia-nitric acid system during winter: Implications for $PM_{2.5}$ control strategies. *J. Geophys. Res.* 110, D07S14.

[24] Paerl H.W., Dennis R.L. and Whitail D.R. (2002). Atmospheric Deposition of Nitrogen: Implications for Nutrient Over-Enrichment of Coastal Waters, *Estuaries* 25(4B), 677.

[25] Yang K., Zhou X.N., Yan W.A., Hang D.R. and Steinmann P. (2008). Landfills in Jiangsu province, China, and potential threats for public health: Leachate appraisal and spatial analysis using geographic information system and remote sensing. *Waste Management* 28, 2750.

[26] Graetzel M. and Frank A. (1982). Interfacial electron-transfer reaction in colloidal semiconductor dispersions: kinetic analysis. *J. Phys. Chem.* 86, 2964.

[27] Lantz J. and Corn R. (1994). Time-resolved optical second harmonic generation measurements of picosecond band flatting processes at single crystal TiO_2 electrodes. *J. Phys. Chem.* 98, 9387.

[28] Chen F., Xie Y., Zhao J. and Lu G. (2001). Photocatalytic degradation of dyes on a magnetically separated photocatalyst under visible and UV irradiation. *Chemosphere* 44, 1159.

[29] Linsebigler A.L., Lu G. and Yates J.T. (1995). Photocatalysis on TiO2 surfaces: principles, mechanisms, and selected results. *Chem. Rev.* 95, 735.

[30] Levy B., Liu W. and Gilbert S. (1997). Direct Photocurrents in Nanostructured TiO_2/SnO_2 Heterojunction Diodes. *J. Phys. Chem. B* 101, 1810.

[31] Shang J., Yao W.Q., Zhu Y.F. and Wu N. (2004). Structure and photocatalytic performances of glass/SnO_2/TiO_2 interface composite film. *Appl Catal A*, 257, 25.

[32] Kolen'ko Y.V., Churagulov B.R., Kunst M., Mazerolles L. and Justin C.C. (2004). Photocatalytic properties of titania powders prepared by hydrothermal method. *Appl Catal B Environ.* 54, 51.

[33] Mo J., Zhang Y., Xu Q., Lamson J.J., Zhao R. (2009). Photocatalytic purification of volatile organic compounds in indoor air: A literature review. *Atmos. Environ.* 43:2229.

[34] Ao C.H., Lee S.C., Yu J.Z., Xu J.H. (2004). Photodegradation of formaldehyde by photocatalyst TiO_2: effects on the presences of NO, SO_2 and VOCs. *Appl. Catal. B Environ.* 54, 41.

[35] Chang C.P., Chen J.N., Lu M.C. (2003). Heterogeneous photocatalytic oxidation of acetone for air purification by near UV-irradiated titanium dioxide. *J. Environ. Sci. and Health Part A – Toxic/Hazardous Substances and Environmental Engineering*, 38, 1131.

[36] Luo Y, Ollis DF, (1996). Heterogeneous photocatalytic oxidation of trichloroethylene and toluene mixtures in air: kinetic promotion and inhibition, time-dependent catalyst activity. *J. Catal.* 163, 1.

[37] Kim S.B., Hong S.C. (2002). Kinetic study for photocatalytic degradation of volatile organic compounds in air using thin film TiO_2 photocatalyst. *Appl. Catal. B Environ.* 35, 305.

[38] Obee T.N., Hay S.O., (1997). Effects of moisture and temperature on the photooxidation of ethylene on titania. *Environ. Sci. Technol.* 31, 2034.

[39] Miller R. and Fox R. (1993). in: *Photocatalytic Purification and Treatment of Water and Air*, eds. D. F. Ollis and H. Al-Ekabi (Elsevier, Amsterdam) 573.

[40] Batzill M. and Diebold U. (2005). The surface and materials science of tin oxide. *Prog. Surf. Sci.* 79, 47.

[41] Vinodgopal K., Bedja I., Kamat P.V. (1996). Nanostructured Semiconductor Films for Photocatalysis. Photoelectrochemical Behavior of SnO_2/TiO_2 Composite Systems and Its Role in Photocatalytic Degradation of a Textile Azo Dye. *Chem. Mater.* 8, 2180.

[42] Solis-Casadors D, Vigueras-Santiago E, Hernandez-Lopez S, Camacho-Lopez MA (2009). Characterization and Photocatalytic Performance of Tin Oxide. *Ind. Eng. Chem. Res.* 48, 1249.

[43] Zhan P.I., Li C., Zhong J., Gong M., Chen Y. (2008). Gas-Phase Photocatalytic Oxidation of Benzene over TiO_2 Supported on $Ce_{0.5\text{-}x}Zr_{0.5\text{-}x}Ba_{2x}O_2$. *Chinese Journal of Catalysis*, 29, 453.

[44] Nauman E.B. (2002). *Chemical Reactor Design, Optimization and Scaleup*, McGraw-Hill, New York, 185.

[45] Liu H., Lian Z., Ye X. and Shangguan W. (2005). Kinetic analysis of photocatalytic oxidation of gas-phase formaldehyde over titanium dioxide, *Chemosphere* 60, 630.

[46] Chin P., Yang L.P.and Ollis D.F. (2006). Formaldehyde removal from air via a rotating adsorbent combined with a photocatalyst reactor: Kinetic modeling, *J. Catal.* 237, 29.

[47] Hossain M.M., Raupp G.B., Hay S.O. and Obee T.N. (1999). Three-dimensional developing flow model for photocatalytic monolith reactors, *AIChE J.* 45, 1309.

[48] Bakardjieva S., Subrt J., Stengl V., Dianez M.J., Sayagues M.J. (2005). Photoactivity of anatase-rutile TiO_2 nanocrystalline mixtures obtained by heat treatment of homogeneously precipitated anatase, *Appl. Catal. B: Environ.* 58, 193.

[49] He J.B., Zhang X., Wei F.Y., Chen P.R. (1999). Effect of Crystallization phases of titania film on photocatalytic degradation of aniline, *Chinese Journal of Applied Chemistry*, 16(5), 57.

INDEX

A

B

C